Transnational Geographies of the Heart

T0176723

RGS-IBG Book Series

For further information about the series and a full list of published and forthcoming titles please visit www.rgsbookseries.com

Published

Transnational Geographies of the Heart

Intimate Subjectivities in a Globalising City

Katie Walsh

WILEY Blackwell

This edition first published 2018
© 2018 John Wiley & Sons Ltd

All rights reserved. No part of this publication may be reproduced, stored in a retrieval system, or transmitted, in any form or by any means, electronic, mechanical, photocopying, recording or otherwise, except as permitted by law. Advice on how to obtain permission to reuse material from this title is available at http://www.wiley.com/go/permissions.

The right of Katie Walsh to be identified as the author of this work has been asserted in accordance with law.

Registered Office(s)
John Wiley & Sons, Inc., 111 River Street, Hoboken, NJ 07030, USA
John Wiley & Sons Ltd, The Atrium, Southern Gate, Chichester, West Sussex, PO19 8SQ, UK

Editorial Office
9600 Garsington Road, Oxford, OX4 2DQ, UK

For details of our global editorial offices, customer services, and more information about Wiley products visit us at www.wiley.com.

Wiley also publishes its books in a variety of electronic formats and by print-on-demand. Some content that appears in standard print versions of this book may not be available in other formats.

Limit of Liability/Disclaimer of Warranty
While the publisher and authors have used their best efforts in preparing this work, they make no representations or warranties with respect to the accuracy or completeness of the contents of this work and specifically disclaim all warranties, including without limitation any implied warranties of merchantability or fitness for a particular purpose. No warranty may be created or extended by sales representatives, written sales materials or promotional statements for this work. The fact that an organization, website, or product is referred to in this work as a citation and/or potential source of further information does not mean that the publisher and authors endorse the information or services the organization, website, or product may provide or recommendations it may make. This work is sold with the understanding that the publisher is not engaged in rendering professional services. The advice and strategies contained herein may not be suitable for your situation. You should consult with a specialist where appropriate. Further, readers should be aware that websites listed in this work may have changed or disappeared between when this work was written and when it is read. Neither the publisher nor authors shall be liable for any loss of profit or any other commercial damages, including but not limited to special, incidental, consequential, or other damages.

Library of Congress Cataloging-in-Publication Data
Names: Walsh, Katie, author.
Title: Transnational geographies of the heart : intimate subjectivities in a globalising city / by Katie Walsh.
Description: Hoboken, NJ : John Wiley & Sons, 2018. | Series: RGS-IBG book series | Includes bibliographical references and index.
Identifiers: LCCN 2017053857 (print) | LCCN 2018004091 (ebook) | ISBN 9781119050438 (pdf) | ISBN 9781119050421 (epub) | ISBN 9781119050452 (cloth) | ISBN 9781119050445 (pbk.)
Subjects: LCSH: British–United Arab Emirates–Social life and customs. | Interpersonal relations. | Aliens–United Arab Emirates.
Classification: LCC DS219.B75 (ebook) | LCC DS219.B75 W35 2018 (print) | DDC 305.82/105357–dc23
LC record available at https://lccn.loc.gov/2017053857

Cover Design: Wiley
Cover Image: © udra/Gettyimages

Set in 10/12pt Plantin by SPi Global, Pondicherry, India

The information, practices and views in this book are those of the author(s) and do not necessarily reflect the opinion of the Royal Geographical Society (with IBG).

Printed and bound by CPI Group (UK) Ltd, Croydon, CR0 4YY

10 9 8 7 6 5 4 3 2 1

Contents

Series Editor's Preface

The RGS-IBG Book Series only publishes work of the highest international standing. Its emphasis is on distinctive new developments in human and physical geography, although it is also open to contributions from cognate disciplines whose interests overlap with those of geographers. The Series places strong emphasis on theoretically informed and empirically strong texts. Reflecting the vibrant and diverse theoretical and empirical agendas that characterise the contemporary discipline, contributions are expected to inform, challenge and stimulate the reader. Overall, the RGS-IBG Book Series seeks to promote scholarly publications that leave an intellectual mark and change the way readers think about particular issues, methods or theories.

For details on how to submit a proposal please visit:
www.rgsbookseries.com

David Featherstone
University of Glasgow, UK

RGS-IBG Book Series Editor

Acknowledgements

The original research on which this book is based was funded by an ESRC Postgraduate Training Studentship (R42200134499).

I am enormously thankful to all the people in Dubai who shared their insights and everyday lives with me as interviewees in the research. I cannot thank you individually, unfortunately, or I would risk your anonymity. I hope that I have done enough to demonstrate both the diversity of perspectives and experiences I encountered, as well as the vulnerabilities of British residents in Dubai individually and collectively. In this book, I have tried to locate myself, however temporarily, as part of your community, not least because I felt a strong sense of belonging to this moment in Dubai, among you, and I am enormously grateful for the friendships I experienced, many of which were life-transforming.

The idea for the ethnographic research on which this book is based emerged during my year as an MA student (2000–2001) on the Cultural Geography (Research) PGT programme at Royal Holloway, University of London. I am indebted to the intellectual environment I encountered there, especially to Philip Crang who supervised my PhD, Katie Willis who mentored my postdoctoral year, and Catherine Nash and David Gilbert who inspired and supported me at various points during my MA year and beyond. I am also thankful to Alison Blunt and Philip Jackson whose feedback on my fieldwork as examiners gave me the confidence to highlight my contribution all these years later. While studying at RHUL I also had the great fortune to be supported by the friendship of fellow students, especially Becky Fox, Fernando Garcia and Hilary Geoghegan. My mother and father deserve special mention, too, not least for providing a 'boomerang' home for me in the last few months of writing up my research and again before I started teaching at Sussex.

This book is also a product of reflection in the decade following my PhD. Meeting and collaborating with other people writing on migration and/or British migrants has been especially rewarding and the insights of Anne-Meike Fechter, Anne Coles, Pauline Leonard and Karen O'Reilly have been especially helpful.

I am lucky also to have worked for ten years in the Department of Geography at University of Sussex and in affiliation with the Sussex Centre for Migration Research. Here I have had the privilege of learning both from those who have already shaped the field of Migration Studies – especially Russell King – as well as from doctoral students I have had the delight to co-supervise and whose work has shaped my own. I have appreciated collegial support and encouragement from many others at Sussex and I am especially grateful to those who collaborated with running field classes to Dubai during which my enthusiasm for this monograph was re-booted.

This book draws upon empirical material I have previously published in the following articles and book chapters (Coles & Walsh 2010; Walsh 2006a, 2006b, 2007, 2008, 2009, 2010, 2011, 2012).

Chapter One
Introduction

In late 2002, I flew from London Heathrow to Dubai to join the approximately 60,000 other British nationals living in the United Arab Emirates at the time (Sriskandarajah & Drew 2006). So, 'Why Dubai?', I was frequently asked, by academics and British migrants alike. My attempt to explain why I had chosen this city as the site of my fieldwork mixed convoluted justifications of an academic kind with confessions of searching the internet to find information about a region of which I had previously been completely ignorant. Yet, for someone intrigued by migrant subjectivities, temporarily relocating there made sense: Dubai was, shortly afterwards, acknowledged as among the 'most global' of cities with over one million 'foreign born' residents (Price & Benton-Short 2007). Large and diverse migrant populations, especially from south Asia, were already present by the millennium, yet academic engagement with migration to the Gulf Cooperation Council (GCC) countries, at least by those writing in English, was still extremely limited. The United Arab Emirates, and especially Dubai, were undergoing a period of rapid economic transformation, accompanied by a 'super-fast urbanism' (Bagaeen 2007), and were showcasing a new mode of globalisation that would be replicated across the region and that continues to resonate with transformations in city-building across the world (Elsheshtawy 2010). In 2002, then, Dubai was not yet the global city it is today, at least not in traditional terms, but it was a rapidly *globalising city* nonetheless (see Yeoh 1999). As such, Dubai was a productive site through which to explore transnational migrants' 'intimate subjectivities' (Constable 2016; Mahdavi 2016) as they lived the global–local intersections of this newly unfolding postcolonial urbanity. Attention to the emplaced, embodied, and emotional production and negotiation of

Transnational Geographies of the Heart: Intimate Subjectivities in a Globalising City, First Edition. Katie Walsh.
© 2018 John Wiley & Sons Ltd. Published 2018 by John Wiley & Sons Ltd.

intimacy, I will suggest, responds to wider calls to examine the 'stickier' moments of migrants' everyday lives, illuminating the processes of reterritorialisation in transnational spaces (Jackson, Crang & Dwyer 2004).

The United Arab Emirates (UAE) is also a major destination for British migrants, equal in 10th place with Switzerland in terms of the number of UK nationals living abroad (Finch 2010)[1]. This is all the more remarkable since the nine countries that have more UK nationals living permanently abroad are either located within the European Union – where mobility for British nationals is at the time of writing still relatively straightforward (in the case of Spain, France, Ireland and Germany) – or are Anglophone former British settler colonies (Australia, USA, Canada, New Zealand and South Africa) where we might anticipate long-established patterns of migration to be maintained (see Finch 2010: 29). The British, and privileged migrants more generally, are frequently overlooked in mainstream policy and academic migration debates, especially in media discourses in the UK which almost entirely ignore out-migration. Most Britons resident in the UAE are living in Dubai or Abu Dhabi and are either there to work themselves or to accompany a spouse or parent. Indeed, the *Kafala* sponsorship system, discussed in more detail in Chapter 3, determines that residency in the UAE (as across the GCC countries more generally) is dependent on the sponsorship of an employer or a close family member.

It was only in the late 1960s that, following the discovery of oil and the political independence of the region, economic and infrastructural development started to bring Britons to live in what was, until independence in 1971, a British protectorate: the 'Trucial States' (Coles & Walsh 2010). Dubai's emergence on the global stage as an 'instant city' (Bagaeen 2007) from 2000 onwards negates the much longer histories of the region, including the ongoing implications of British imperial involvement. The initial communities of British residents remained very small, for example, in 1968, the first census in Dubai recorded only 400 Britons a highly skilled professional migrant group consisting of a handful of advisors to the government, as well as the managerial level staff of banks and trading, shipping, and oil companies and a few teachers, health professionals, and town planners (Coles & Walsh 2010). By the mid-1980s, however, Findlay (1988) identified the Middle East, with the Gulf countries of notable significance, as emerging ahead of the Old Commonwealth countries as a destination for professional and managerial level migrants from the UK. Though British migrants in Dubai are hugely outnumbered by the much larger south Asian communities resident in the city, an analysis of this particular 'postcolonial' positionality, and their reproduction of privileged 'expatriate' subjectivities marked by nationality, class and race, is an important part of understanding Gulf subjectivities and global mobilities more broadly.

In contrast, the same question, 'Why Dubai?', when directed towards the British migrants I had temporarily relocated to interview, received rapid and seemingly straightforward replies: 'It's a tax-free sunshine.' This response would not be a surprise to anyone who has spent time with British migrants on a personal or research basis. The significance of both income and lifestyle in shaping global flows of privileged migration are well established in a wider interdisciplinary literature (e.g. Knowles & Harper 2009). Indeed, this literature has critically examined the relative privilege of this group as something that migration often amplifies, especially in terms of the racialisation of expatriate identities through whiteness (e.g. Leonard 2010). The phrase 'tax-free sunshine' also illuminates the economic diversification strategies that, at the time of my fieldwork in the early 2000s, were transforming Dubai into the global city it has become today. One of these centred on the provision of themed free-zones (for instance, Media City, Internet City, Education City, Healthcare City) set up to attract companies through the provision of economic incentives such as the allowance of tax-free salaries to be paid to expatriate employees (Davidson 2008). Dubai's government has also actively sought to provide leisure and consumption opportunities to encourage a wealthy, highly skilled, and aspirational middle class to *temporarily* relocate from across the world. Arguably, Dubai has been hugely successful in this aim. Of course, its super-diversity is structured by massive inequalities and, as such, academics have critiqued its migrant labour regime (e.g. Buckley 2012; Davis 2005). From the perspective of British migrants, however, even non-graduates and those with a more modest skill set, 'tax-free' equates to more career opportunities and a higher disposable income than they could achieve at home, with the possibility of consuming an elite lifestyle. Meanwhile, highly skilled inter-company transfers and more traditional 'expatriate packages' in which accommodation, private healthcare, schooling and flights are added extras enable most families to move on the salary of one household member. This 'lead' migrant, whose career trajectory determines household mobility, is usually male so, for some couples at least, relocation to Dubai marks a shift from a more gender-neutral household structure based on the dual-career or dual-income strategies that have become increasingly common among middle-income couples in the UK.

Many of my interviewees suspected my interest in their lives had arisen from an 'expat' childhood, but I had never previously lived abroad. I was there to research their sense of home and belonging, prompted instead by questions about materialities and diaspora that had emerged, for me at least, from an academic curiosity fuelled by transformations in geography and cultural studies from the late 1990s. This book is still about those things, but it also departs massively from this primary objective of the fieldwork to consider the questions, narratives and observations that emerged unbidden during my fieldwork and invited me to explore their significance. Domesticity, transnationalism, belonging, home and identity, continue to be threads of analysis that run through this volume – they weave through the production of our intimate selves after all – but my focus is more directly on our geographies of the heart. Intimacy, highlighting personal relationships and the array of closer connections through which British migrants negotiate belonging in everyday life, turned out to dominate other people's telling of life in Dubai and my listening. I found myself increasingly exploring the textures of intimacy negotiated and enacted by British migrants in this particular spatial-temporal, and thoroughly transnational, urban site, but without a language for doing so.

Fortunately, geography, migration studies, and the wider social sciences have since been on their own theoretical journeys, providing me with the building blocks to more thoroughly explore these transnational geographies of the heart that I observed. With this analytical focus, I join a number of other scholars concerned with the embodied and emotional dimensions of mobilities (e.g. Boccagni & Baldassar 2015; Conradson & McKay 2007; Dunn 2009; Mai & King 2009) and with the links between globalisation and our intimate lives (e.g. Baldassar & Merla 2014; Beck & Beck-Gernsheim 2014; Padilla et al. 2007; Pratt & Rosner 2012). Accounts of migrants' 'intimate subjectivities' more specifically have largely been conducted in contexts whereby the global commodification of intimacy is more overt – in domestic work, sex work and cross-border marriages, especially – illuminating the power relations involved in cross-cultural intimacies, the restrictions on migrant women's reproductive rights arising from their marginalised status as contracted labour migrants, and the challenges of transnational family life (e.g. Constable 2009, 2016; Mahdavi 2016; Pratt 2012). Nonetheless, while I am focusing on freely established relationships from the perspective of relatively privileged migrants, questions of power and the impact of global work on the reproduction of migrants' subjectivities remain important. In the analysis, I reveal the everyday efforts of the 'doing' of our intimate selves, with attention to how gender, class, nationality and race intersect. I argue that British migrants in Dubai enact intimate subjectivities marked by privilege (in terms of being middle class, white, and heterosexual), but also illuminate the diversity and instabilities of their social locations, the ambiguous impact of their migrant status, the ongoing work involved in the reconstitution of heteronormativity and the interplay of gendered subjectivities.

Existing literatures on intimate subjectivities have tended to privilege one kind of relationship, such as parenting or romance, and often, as a result, one kind of practice of intimacy (especially care). Here I depart from this emphasis to explore instead a range of interpersonal relationships that all inform migrants' intimate subjectivities, including friendship, 'community', and couple relations (both marital and short-term sexual encounters). This is important because the meaning of each of these kinds of relationship is not entirely distinct, but also because our experiences of each kind of relationship are not emotionally separated from one another in everyday life. Put another way, an individual's experience of couple or 'love' ties cannot be isolated from their enactment of friendship or family. Our understanding of the collective norms, shifts, and practices of each of these different kinds of personal relationships in migration thereby benefits from exploring the intersections of intimate subjectivities.

In this book, then, I examine migrants' intimate subjectivities through ethnographic attention to different spaces, practices and accounts of intimacy in the globalising city of Dubai. In doing so, I argue for the significance of *geogra-phies* of intimacy: this is about an understanding of the spatialisation of intimate subjectivities and the importance of place in their ongoing constitution. I show how multiple sites of belonging work to shape our interpersonal relationships: bodies, homes, city spaces and transnational spaces. For example, the specific racialisation of the city helps to regulate intimacy, serving to entrench the signif-icance of nationality in friendships and families; but these textures of intimacy are shaped also by transnational flows of people, objects and media, as well as by British migrants' embodied performances of 'expatriate' subjectivities. Thereby, we see the significance of a geographical analysis in understanding this multi-plicity in how intimacy is lived and imagined. Intimacy, I will argue, is one of the most significant, yet unacknowledged, discursive reference points in migrants' narratives of their belonging and in their everyday navigation of a dialectic of home and migration.

British Global Mobilities

In their study of British migration (Finch 2010), the Institute of Public Policy Research estimated that some 5.6 million UK nationals were living outside the UK in 2008. This extraordinary statistic – equivalent to one in ten Britons – encompasses migrants of varying purpose and permanence, with diverse practices of settlement and mobility. A number of studies of British migration have been conducted and framed by a literature on highly skilled migration. For example, Beaverstock's (1996, 2005) study of inter-company transferees in New York focused on highly skilled migrants relocating within and between financial transnational corporations. Beaverstock (2005: 246) identified these British resi-dents as part of a transnational elite, 'nomadic' workers 'whose ultimate interna-tional mobility meets the challenges of international business in globalisation'. Typically they were male (80%), married (60%) and were posted to New York for two years. Their career paths were often already international, however, with 40% having worked and lived in other financial centres for over a year, and with reg-ular business travel maintaining these transnational connections (Beaverstock 2005). Their travel was highly privileged, with expatriate packages including adjusted salaries, generous housing allowances, business flights, relocation payments, and subsidies for school fees, health and personal insurance. These British migrants, like the scientists interviewed by Harvey (2008) in Boston's pharmaceutical and biotechnology sector, are a valued part of the knowledge economy. Indeed, for Beaverstock (2011), 'expatriation' is an important process for world cities to secure 'talented' human capital and he positions British migration as part of this global process. In this way British migrants are figured as part of the 'transnational capitalist class' (Sklair 2000), defined by their global connections and mobilities, agents of a globalising world defined by movement.

(Highly) skilled migration is widely recognised as a strongly gendered process, with the accompanying spouse usually female (Coles & Fechter 2008; Hardill 1998; Hardill & MacDonald 1998; Kofman & Raghuram 2005).

This depiction of British migrants resonates with the professional occupation and privileged global status of many of the British residents I encountered in Dubai, but it is only part of the story. Existing studies of British migrants in global cities have consistently demonstrated that 'expatriate communities' are rarely homogenous and are increasingly stratified by class (as indicated by occupation, affluence, consumption practices and education), age, gender, marital status and length of stay (Knowles & Harper 2009; Leonard 2010; Scott 2006). Further evidence of this stratification is provided by ethnographic studies of 'Western' migrants (Farrer 2010; Fechter 2007). Such studies have focused mainly on cities in Asia, especially Hong Kong (Knowles & Harper 2009; Leonard 2010). In some cases, this research has shown that single working men and, increasingly, women disrupt the heteronormative picture of skilled 'family' migration, making the production of gendered migrant subjectivities more diverse (e.g. Fechter 2008; Willis & Yeoh 2008)[2]. Other research has shown that British migrants' identities in global cities may be further differentiated by lifestyle factors (e.g. Knowles & Harper 2009; Scott 2007). While their perspective on the 'good life' may be constituted rather differently, their prioritisation of 'lifestyle' resonates with studies of Northern European migration to rural and coastal areas of Southern Europe (e.g. Benson 2010; King, Warnes & Williams 2000; Oliver 2008; O'Reilly 2000). For some, of course, these economic and lifestyle factors go hand in hand, intertwined in accounts of disposable income and class mobility, and Knowles and Harper (2009: 11) remind us that the recognition of lifestyle factors as significant need not lead to us deny that Britons are also migrating for work and/or a career: 'Our use of lifestyle migration also acknowledges the inseparability of economic factors like income, and the quality of life it supports. In the global scheme of things, those who are already economically privileged are able to prioritise qualities beyond life's financial elements.'

The role of mobility in the making of the middle-class self is explored more fully by Benson and O'Reilly (2009) in their articulation of the meaning of lifestyle migration. For them, lifestyle migration is 'a search, a project, rather than an act, and it encompasses diverse destinations, desires and dreams' (Benson and O'Reilly 2009: 610). As such, it is 'necessarily comparative' (ibid.: 610) and 'part of their reflexive project of the self' (ibid.: 615, drawing on Giddens). As Scott (2007: 1123) explained with reference to British migrants in Paris, mobility is 'now a dominant feature of middle class reproduction', such that not only migration itself but also the choice of location 'can help to confer status by distinguishing them in a socio-geographical sense from other middle class groups' (Scott 2007: 1124). Conradson and Latham's (2005) notion of 'middling transnationalism' remains salient, then, in terms of describing the status of most British migrants, even while the heterogeneity of their backgrounds, motivations and social locations are just beginning to be understood. In spite of their relatively privileged status in comparison with many other migrant workers in Dubai, British residents still do not have the right to apply for UAE citizenship. While some have recently gained access to a more permanent form of visa through property ownership or investment in business, most Britons gain residency in Dubai on a biannual basis through the sponsorship of their employer (or as a dependent on the visa of their spouse or parent). Indeed, the *Kafala* sponsorship system and its impact upon intimacies is discussed further in Chapter 3 and then traced through the book. At this point, however, it is relevant to highlight that one impact of their residence being tied to short-term contracts is that a sense of 'temporariness' often frames the experiences of Britons in the Gulf, irrespective of their actual length of residence. As a result, it is also useful and appropriate to consider the British through the theoretical lens of transnationalism, since they often continue to orientate their sense of belonging to the UK as well as Dubai. Routine 'transnational practices' (Vertovec 2001) such as property investments and the maintenance of interpersonal relationships with family and friends through annual visits and the use of ICTs (Information and Communication Technologies), are therefore relevant to any analysis of their everyday lives. Nonetheless, their 'emplacement' in Dubai is also significant, as described in other recent explications of skilled migration (e.g. Meier 2014; Riemsdijk 2014; Ryan & Mulholland 2014). The relative privilege of British migrants in terms of being able to negotiate this mobility–settlement dialectic in terms that suit them is captured by the term 'expatriate'. This is a term I further explore in the next section.

Postcolonial Histories, Cities, Migrations

Given the relative privilege of British migrants, an analysis of their intimate subjectivities might not seem especially useful or necessary. The location of this study – Dubai – would perhaps further discourage engaging in such an analysis, since of more urgent concern are the impacts of the Gulf Cooperation Council (GCC) countries' migration regimes on the human rights of low-income migrant workers, especially construction workers from south Asia (e.g. Human Rights Watch 2006), and questions about the environmental and social sustainability of the modes of urbanisation being produced (e.g. Davis 2005). Yet, the experiences of low- and middle-income south Asians across the GCC countries are now much better understood, resulting from a recent interdisciplinary literature on Gulf migration (e.g. Buckley 2012; Elsheshtawy 2010; Gardner 2010; Kamrava & Babar 2009; Mahdavi 2016; Osella & Osella 2007; Vora 2013). A focus here on relatively affluent, or at least middle-class, migrants in the Gulf, especially those who are racialised through whiteness, complements and complicates this emerging literature. In this section, I make a case for selecting British migrants as my foci for research on Dubai migrant subjectivities, by drawing on postcolonial theoretical perspectives. Scholars of postcolonial theory emphasise that the 'post' should not be used to refer to a specific historical moment, political status, or spatial context of formal decolonisation, rather that postcolonialism might be better understood as a productive set of critical approaches to colonialism and its legacies (Hall 1996). With this in mind, I suggest three reasons why we might be attentive to postcolonial perspectives: firstly, I provide a discussion, albeit brief, of Britain's informal empire in the Gulf, 1820–1971, focusing on the 'protectorate' relationship with the 'Trucial States' (Onley 2005). Secondly, I consider the application of a 'postcolonial cities' (Yeoh 2001) perspective to Dubai, in order to acknowledge the traces of colonial ontologies of difference that shape the stratification of the globalising city through notions of race, ethnicity and nationality. Thirdly, I consider global British migration itself as one among many differently navigated 'postcolonial migrations' (Mains et al. 2013), looking at how 'expatriate' subjectivities are racialised through whiteness (Fechter & Walsh 2012; Knowles & Harper 2009; Leonard 2010). The empirical chapters provide evidence for the significance of all three perspectives in exploring intimacy in the globalising city of Dubai in the early to mid-2000s.

Firstly, then, a postcolonial perspective forces us to examine the production of Gulf subjectivities with an acknowledgement of the histories of this region. One of the main issues with much existing Western commentary on Dubai is the way it wipes away any acknowledgement of historical settlement on the Arabian peninsula. Consider, for instance, how Krane (2009) introduces the United Arab Emirates:

> The Arabian Peninsula is a sun-hammered land of drifting sands and rubble wastes. Ranges of unnamed peaks slash across the landscape their sun-shattered rock sharp enough to cut skin. Salt flats shimmer in the moonlight night after night, untouched by humans for eternity... the United Arab Emirates sit on the southeastern corner of Arabia, the most desolate corner of a desolate land... History simply happened elsewhere... (Krane 2009: 3–5).

Locating this hyper-modern city firmly back in a 'timeless' desert is, arguably, a strategic representational move that allows 'Western' commentators and their audience to assume an ethnocentric superiority in their evaluative standpoint. The presumed complete lack of history in Dubai is an idea that circulates in global media and local 'expatriate' discourses alike, yet is disputed by scholarly analysis of the region, including archaeological and historical accounts of the pre-Islamic period, the Islamic period, and the tribal society through which the traditional economies, such as fishing and pearling, were organised (Heard-Bey 2001, 2004; see also Al Abed & Hellyer 2001). Had the region been empty desert, it would not have attracted the attention of the Portuguese in the sixteenth century: they were the first European imperial power to attempt to gain control of parts of what was then known in Europe as 'Historic Oman'. Their motivation was primarily commercial – to control the spice trade with this vital trading port between Africa, Asia and the European markets. As Heard-Bey (2004: 272) argues:

> ... it was not just a group of daring Portuguese adventurers who conquered the traditional trade emporia of the Gulf and parts of the Indian ocean; this was the result of carefully prepared strategy at the Court in Lisbon aimed at taking over by any means possible every sector of the very profitable trade between the Indian Ocean coasts and Europe.

Though the Portuguese were expelled from all the ports by the mid-seventeenth century, their presence created conflict with English and Dutch trading companies who wished to replace them, as well as with Persians trying to promote their sovereignty, such that a new Arab power was able to emerge: the Qawāsim (Heard-Bey 2004). The Qawāsim Sheikhs ruled from Ras al Khaimah over much of the northern coast of the Arabian Peninsula. British ships using the Strait of Hormuz refused to pay the tolls requested by the Qawāsim fleet, which responded with piracy (Onley 2005).

As a result, authorities in British India sent a naval expedition to impose an anti-piracy treaty on all the rulers and governors of the Coast of Oman and, from 1820, a Political Agent was headquartered on Qishm Island (Onley 2005), now a popular stopping point for tourist *dhows*. Onley (2005) describes how this post was amalgamated in the new role of Resident in the Persian Gulf who became responsible for Britain's relations with the entire Gulf region, supported by a naval squadron to patrol the Gulf waters, and how, under this control from British India, a series of treaties or 'Exclusive Agreements' were imposed over the next 150 years (also signed by Bahrain, Qatar and Kuwait). While imperial historians are divided as to whether they view the Gulf states as part of Britain's formal or informal empire during this time (see Onley 2005), either way, British influence was maintained and the treaties served British economic and political interests. As Heard-Bey (2001: 117) explains:

> The UAE never was a colony, but its forerunner, the 'Trucial States', was increasingly absorbed into the British orbit by a system of agreements which successive British governments, first in Delhi and then in London, deemed necessary in order to best pursue their particular objectives of the day.

In his life story, *From Rags to Riches*, Mohammed Al-Fahim (1995) is one of the few commentators to present an Emirati perspective on the historical British presence in this region. He described British imperialism as 'unwanted domination' (ibid.: 27), revealing how political events in Europe and colonial goals in Asia 'had a deep and lasting impact on our development'. When the British destroyed the Qawāsim's fleet of ships and imposed restrictions on the size of new ships, they also destroyed trading and seafaring capabilities, as well as the shipbuilding industry (Al-Fahim 1995: 37). Another treaty stopped the import and export of arms, 'completely disregarding the inhabitants' need to be armed for defensive purposes, as well as for pride and prestige' (ibid.: 35). In addition, concessions were imposed on the pearling industry, ensuring a monopoly for Indian merchants that led to decades of poverty in the region, long before the industry faced competition from Japanese cultured pearls (ibid.: 40–41).

From the 1930s, British policy towards the Trucial States became much more intrusive, in preparation for the operation of British oil company personnel, and the way in which the rulers dealt with domestic matters, such as intertribal strife, was more closely monitored (Heard-Bey 2001). In 1955, for example, the British, knowing the oil-producing potential of the Buraimi area at the intersection of Saudi Arabia, Oman and the UAE, violently cleared Saudi troops to ensure American oil companies would not secure the concession (Al-Fahim 1995: 62). Al-Fahim (1995: 73) describes the development of the oil industry as one of 'mixed blessings'. Prior to independence, he argues:

> The oil companies discouraged the locals from participating in any way other than as hired hands. The people who worked for the oil companies during the 1950s and early 1960s look back with bitterness rather than fondness on those early days. Many have since become prominent citizens; some are ministers or ambassadors, others are highly qualified people in their respective professions. The painful memories of their exploitation, however, linger on even today. While they may have forgiven, they can never forget (Al-Fahim 1995: 73).

While locals were employed as manual labourers, they were supervised by expatriates from India and neighbouring Arab countries (Syria, Lebanon and Palestine), such that: 'The promise of the oil industry proved empty for Abu Dhabians, at least in the early stages. They were under-paid, treated poorly, denied opportunities and made to work in very harsh conditions' (al-Fahim 1995: 76). In 1966, when Sheikh Zayed became the Ruler of Abu Dhabi, he began to renegotiate the concessions and push for an improvement in terms and conditions for employees (Al-Fahim 1995). Today, the government of Abu Dhabi holds the controlling interest in the Emirate's oil industry.

For Stephenson (2013: 7), Dubai's urban expansion since 2000, funded initially by oil-related income, the spectacular mega-projects that others scorn so readily, 'not only stands as a representation of symbolic capital but also as a testimony to decolonisation, signifying ways in which Dubai (and the UAE) has moved towards an agenda of self-determination, political autonomy and economic freedom'. Kanna (2014) too sees echoes of Dubai's historical position as a node in British imperial networks at play in its contemporary urbanscape, especially in its privatised leisure spaces predicated on exclusivism, and argues that:

> Dubai elites – the ruling family above all, but also allies of the ruling family in the merchant class – drew upon British patronage and Western ideologies of free trade, consumption, and racial management to help produce, or (as Chopra might say) coproduce, urban space in post-independence Dubai (Kanna 2014, 606).

Kanna (2014: 607), influenced by Henri Lefebvre's notion of space as a 'palimpsest' that 'archives social struggles both past and present', sees Dubai's current role as a global leisure and tax-free haven as related to its history as a point of political stability in the region. Furthermore, he provides evidence from his interviews that Emiratis are, today, ambivalent towards this urbanisation, raising concerns about the disrespect of Emirati Muslim customs and their feelings of being 'colonised' and excluded from their own country by the presence of Europeans, including the 'sensorial assault' of immodest dress in shopping malls (ibid.: 612). His interlocutors described Europeans as coming from '*al bilad ma bihad al-karama* [countries without a sense of dignity],' leading Kanna (2014: 612) to suggest there is 'a tangible sense among both Emiratis and other non-Europeans of the presence of a neo-colonial hierarchy which privileges European and North Americans'.

In previous publications, most notably in a collaborative paper with Anne Coles (see Coles & Walsh 2010; also Walsh 2010, 2012), I have also argued that it is productive to examine contemporary Dubai using a postcolonial theoretical lens, particularly in terms of the negotiation of its everyday cultural politics and the production of racialised subjectivities. Brenda Yeoh draws together literature on postcolonial cities to identify that:

> The postcolonial city traces continuity rather than disjuncture from its colonial predecessor in the nature and quality of social encounters, which are shot through with notions of 'race' and 'culture' as markers of difference and bases for interaction (Yeoh 2001: 460).

This certainly resonates with urban encounters in Dubai, as will become evident in Chapter 4. The incorporation of 'colonial categories' of difference and privilege (Yeoh 2001, drawing on Jacobs) is also evident within the urban fabric itself, through the government's residential zoning policies, and with respect to the organisation of the labour market. Dubai's labour market is strongly segmented, with the recruitment and employment of migrant labour occurring along distinctly nationalised, ethnicised and racialised lines (see, for example, Malecki & Ewers 2007). The legacies of empire globally, as well as the specific historical relationship between Britain and 'the Trucial States', means that, while they might dispute the allegations of their neo-colonial relationship with Dubai and instead highlight their vulnerabilities as non-citizens subject to the regulatory framework of the *Kafala* sponsorship system (see Chapter 4), contemporary British migrants are relatively privileged in their migrations to Dubai.

Finally, then, and linked to the above, an engagement with postcolonial theory is also relevant to this study because it examines one particular category of 'postcolonial migrant' (Mains et al. 2013): the privileged 'expatriate'. As Alan Lester (2012: 7) argues, 'expatriate' migrants, 'very often actively perform routines and rhythms inscribed through colonial practice' and a number of researchers across disciplines have explored how today's 'expatriate' migrations draw on colonial histories in terms of the routes, discourses and narratives of their everyday lives (e.g. Conway & Leonard 2014; Cranston 2016; Fechter & Walsh 2010, 2012; Knowles 2005; Knowles & Harper 2009; Leonard 2010). Caroline Knowles (2005: 107) convincingly argued that we can find clues to the contemporary salience of empire in the global mobilities of today's British migrants: 'Empire survives as a feeling of choice and opportunity, (divergent) forms of entitlement, facilitated by a (racialised) geography of routes already carved out and traversed by others.' Indeed, the baggage of Empire is even present in the very terminology of 'expatriate' and its usage. Anne-Meike Fechter (Fechter 2007: 1; following Cohen) was among the first to engage critically with the use of 'expatriate' in both colloquial and scholarly contexts, identifying the distinctive application of this terminology to Westerners, in spite of its broader definition from the Latin of 'a person who lives outside their native country,' such that, 'the majority of contemporary migrants who leave their countries to live elsewhere are typically not referred to as expatriates'. This seeming paradox can be explained by the connotations that this term carries with it in terms of the classed and racialised migrant subjectivities it evokes (Fechter & Walsh 2010). This is why Chapter 4 refers to *the making of* 'expatriate' subjectivities through spatialised urban life-worlds and discourses, but why I use the term 'migrant' to refer to my interlocutors throughout (see also Cranston 2016). It is important to note that many Britons do not think of themselves as either migrants or expatriates, and certainly wish to distance themselves from a 'colonialist' and 'imperialist' Britishness. Nevertheless, a postcolonial theoretical perspective is productive in encompassing those who are privileged by colonial legacies, since their movements and subjectivities are a vital part of understanding the racialisation of contemporary global cities.

In spite of the deconstruction of race as a social category, for a long time whiteness remained invisible, normative, unmarked, unnamed and, by extension, a non-racial or racially neutral category, until critical theories of whiteness were explicated (Frankenberg 1993). It is now widely understood that 'race' shapes the lives of those who are privileged by it, not just those whom it oppresses, and that mainstream spatiality is complicit with whiteness (Frankenberg 1993). Recently, the theorisation of 'whiteness' has been used to examine 'postcolonial migrations' in various countries, for example, Indonesia (Fechter 2005) and Hong Kong (Knowles 2005; Leonard 2008, 2010; Yeoh & Willis 2005). Fechter's (2007) ethnography of Westerners in Indonesia, for example, revealed the significance of the marking of raced, classed, and nationalised boundaries in relation to the local community. Race, gender and nationality remain critical to my analysis of intimate subjectivities in this book. In the next section, I explain how this book is situated within a broader literature on transnational migrants' subjectivities.

Theorising Transnationalism: From the Global to the Intimate

At the end of the 1990s, following a decade of analyses of globalisation that uncritically highlighted unbounded mobilities and processes of deterritorialisation, theorists of international migration began to call for micro-analyses that could capture the transnational practices 'embodied in specific social relations established between specific people, situated in unequivocal localities, at historically determined times' (Guarnizo & Smith 1998: 11). Within geography, for example, Kathryn Mitchell (1997a, 1997b) made an appeal against the increasingly abstract depictions of global flows, cautioning against a celebratory understanding of transnationalism as inherently transgressive and questioning the 'hype of hybridity'. In a Special Issue of *Antipode* (Mitchell 1997b), she introduced four 'transnational spatial ethnographies' to locate the specificities of global processes and trace transnational flows through empirical material on 'real' bodies and 'real' geographies. A few years later, Yeoh, Willis and Fakhri (2003: 212) focused on the 'edges' of transnationalism to demonstrate that transnational identities 'while fluid and flexible, are also at the same time grounded in particular places at particular times'. And geographers continued to explore the way in which transnational processes and spaces are 'constituted through the dialectical relations of the grounded and flighty, the settled and the flowing, the sticky and the smooth' (Jackson et al. 2004: 8).

Reflecting on these broader debates a decade later, Geraldine Pratt and Victoria Rosner (2012) outline their theoretical approach – the 'global intimate' – arguing that bringing the global and the intimate into a conceptual relationship can be seen as part of 'a distinctively feminist' response to these questions of transnational relations.[3] The essays in their edited collection 'explicitly adopt a method that involves disrupting the very idea of scale, by sliding between global and intimate, weaving together these two different modes of feminist thought' (ibid.: 19). Pratt and Rosner (2012: 20) argue:

> The intimate directs us to an ethical stance towards the world – namely, an approach that neither simplifies nor stereotypes but is attentive to specific histories and geographies. If we can guard against universalizing particular structures of intimacy and resist any facile connection between women and the intimate, we can find in the idea of the intimate a language and approach that can disrupt tendencies towards totalization that inevitably arise when we try to theorise on a planet-wide basis... By the same token, 'thinking globally' can provide a necessary counterweight to traditional analyses of intimacy by calling attention to the unintended consequences of failing to place the local setting in a broader context or refusing to look at the threads that connect intimate practices to a larger world.

For Pratt and Rosner (2012) then, 'the intimate' is a term that allows us to focus on the more personal dimensions of everyday life, by highlighting our bodies, senses, emotions, affects, attachments, and the materiality of social existence. The rich theoretical possibilities of 'grounding' transnational relations through attention to the 'global intimate' (Pratt & Rosner 2012) and, as part of this, in analysing migrants' intimate subjectivities as I do in this book, reflect broader theoretical shifts in how migration is understood in current literatures. In the remaining discussion of this section, then, I will argue that migration is already conceptualised through a trio of concepts that help us in this task, namely: emplacement, embodiment and emotion.

The global/local dialectic to which Pratt and Rosner (2012) draw attention is also at centre-stage in Michael Smith's (2005: 237) account of 'transnational urbanism', an analysis of contemporary cities that captures 'a sense of distanciated yet situated possibilities for constituting and reconstituting social relations'. His emphasis on translocalities as a way of thinking about 'new modes of being-in the world' (ibid.: 237) highlighted the way in which transnational flows work through local sites. This optic was taken up by Conradson and Latham (2005a: 228) who found the theoretical lens of transnational urbanism attractive precisely for its 'creative incorporation of mobility and emplacement' and the understanding of migrants' everyday lives it therefore affords:

> Whilst acknowledging the scope of contemporary global mobility, transnational urbanism is a concept that remains attentive to the continuing significance of place and locality. In this sense it may be seen as advancing those earlier critical impulses which found first-wave globalisation narratives to be insufficiently nuanced. It eschews accounts of individuals traversing a somehow frictionless world, endorsing instead research that details the emplaced corporealities of such movement (Conradson & Latham 2005a: 228).

Conradson and Latham (2005a: 228) go on to suggest that a focus on the everyday mundane practices involved in transnational mobility is especially productive in this respect, allowing researchers to explore how people negotiate transnational life-worlds and get a sense of the 'texture of the globalising places we inhabit'. As they suggest:

> Viewed from this quotidian angle, even the most hyper-mobile transnational elites are ordinary: they eat; they sleep; they have families who must be raised, educated and taught a set of values. They have friends to keep up with and relatives to honour. While such lives may be stressful and involve significant levels of dislocation, for those in the midst of these patterns of activity, this effort is arguably simply part of the taken-for-granted texture of daily existence. An investigation of the life-worlds of these mobile individuals, and the activities which constitute them, thus provides a useful counterpoint to the inflationary tendencies of some writings on globalisation (Conradson & Latham 2005, 228–289).

Within my own research, I have already contributed to these debates on emplacement by exploring how migrant belonging is constituted across and through the dialectics of migration/home and mobility/settlement in terms of domestic belongings (e.g. Walsh 2006b). Here I extend these debates by exploring how the production of intimate subjectivities operates across these dialectics and informs the texture of everyday space in Dubai.

Alongside these concerns with everyday emplacement of transnationalism, it is possible to trace the emergence of a body of work on migration in which bodies come to the fore and migrants are understood as *embodied* subjects (see also Blunt 2008). Silvey (2005: 144) argues that feminist migration researchers have been at the forefront of these debates, producing research that 'aims to identify and unpack the power relations embedded in, shaped through, and reinforced by migrants' bodies in particular places and across space'. For Silvey (2005) this approach involves 'reclaiming bodies as analytically central and as lived sites of power' (ibid.: 144) and builds on broader feminist contributions theorising embodied subjectivities and the politicisation of identities more generally (ibid.: 142). Migration, then, is revealed as 'a socially embedded process such that it reflects and reinforces social organization' along multiple lines (Silvey 2005: 142). Indeed, feminist migration approaches tend now to examine gender subjectivities as co-produced through race, ethnicity and/or class (e.g. Anthias 2012; Batnitzky, McDowell & Dyer 2008), but with the primary focus on gender nonetheless. The 'gender politics animating migration' (Silvey 2005: 138) have most fully been elucidated in work on the feminisation of labour migration related to the 'commodification of intimacy' (Constable 2009; Ehrenreich & Hochschild 2003), especially domestic workers, sex workers and marriage migrants from Asia navigating the proximities and immobilities of transnational social spaces (e.g. Constable 2003, 2007, 2014; Mahdavi 2011, 2016; McKay 2007; Piper & Roces 2003; Pratt 2012). The language of intimate subjectivities has emerged within this literature, but has not yet been analysed in relation to more privileged migration from the global North or extended to encompass the spatialisation of interpersonal relationships of varying kinds (e.g. friendship and community). It is important to note also that work on migrants' embodiment has not been limited to femininities and there are even some studies focusing on masculinity in the Gulf (see Johnson 2015; Osella & Osella 2000). Of relevance to my own focus on relatively privileged male migrants, the entrenchment of patriarchal social power through an increased economic status brought about by migration has been observed among the transnational business elite (Connell, cited in Shen 2008).

Alongside this gender-focused research, queer-theoretical perspectives have also significantly informed our understanding of migrants as embodied subjects (e.g. Fortier 2003; Gorman-Murray 2007, 2009; Knopp 2004; Luibhéid & Cantú 2005; Manalansan 2006). The 'queer migration' literature has mainly focused on non-heterosexuals, illuminating the mobilities of coming out, the desire to move to a gay neighbourhood or city, and migration for or away from a relationship. But, it might also encompass a range of non-heteronormative others, including straight migrants who are transgressing the norms of citizenship of family life

(Manalansan 2006; Oswin 2010b). For Gorman-Murray (2009: 443–444), for example, 'foregrounding the role of sexuality in migration demands focusing attention on the body itself as it moves through space' and the 'desirous and sensuous dimensions of relocation decisions.'

Finally, emotions have become a much more central and widely adopted framework in migration studies, evident in the recent publication of several special issues that consider various dimensions of the emotional life of migrants, including the emotion work involved in their employment and the management of relationships across transnational fields (e.g. Boccagni & Baldassar 2015; Conradson & McKay 2007; Mai & King 2009; Svašek 2010)[4]. Nicola Mai and Russell King (2009: 297), for example, suggested that the two mainstream research paradigms dominating migration studies were (still) implicitly pushing aside 'the role of emotions, feelings and affect in the motivation and experience of migration'.[5] They identified, firstly, an economic and sociological approach that focuses on the working lives of migrants and the costs and benefits of migration, and a second theoretical agenda, emerging from anthropology and cultural studies, that highlights the exploration of socio-cultural identities and encounters. As a result, they argued:

> emotional relations are regarded as something apart from the economic or the geographic, as something essentially private, removed from the researcher's gaze traditionally fixed on spatial mobility patterns, push-pull factors, the 'laws' of migration, the mobility transition, assimilation/integration and the cross-cultural encounter (Mai & King 2009: 297).

They advocate, then, for an 'emotional turn' in migration and mobility studies which explicitly places emotions at the heart of migration decision-making and behaviour through a focus on love and sexuality. For Mai and King, love and sexuality are concepts that should not be applied only to a distinctive set of migration experiences or to help us understand particular migrant identities but, rather, should be recognised as integral to our understanding of migration as an economic and political process.

To place intimacy at the centre of this analysis, therefore, does not mean that I overlook the economic and structural forces shaping and, indeed, regulating migrants' lives as they participate in global work. Even among privileged and (highly) skilled lifestyle migrants, such as the British in Dubai, it becomes apparent from the empirical chapters later in this book that the emotional dimensions of their lives, including intimacies, are shaped by the materialities in which they are embedded. Sarah Lamb's (2002) early intervention in the debates on transnational ageing highlighted the intimate everyday practices through which transnationalism is constituted. According to Lamb (2002: 300), similarly to Pratt and Rosner (2012), we need to recognise that transnationalism 'involves not only the macro, depersonalised flows of global capital, mass media images and proliferating technologies but also the "intimate", lived everyday lives of particular people'. Lamb (2002) used the term 'intimacy' to signal the everyday embodied practices between kin, especially forms of expressing love between ageing parents and their children. She revealed how something as seemingly 'trivial' as making tea was entangled in the emotional negotiation of familial relationships across generations, as well as the changing meaning of practices in transnational life-worlds.

In arguing for the incorporation of emotions in an understanding of translocal subjectivities, David Conradson and Deidre McKay (2007) argue that it is necessary to start with an understanding of the self as a *relational* achievement. They argue:

> From this perspective who we are derives in part from the multiple connections we have to other people, events and things, whether these are geographically close or distant, located in the present or past. This constellation of others may influence us in diverse ways, acting via physical encounter and somatic internalisation, in response to the power of images and narratives, and through the operation of memory and desire. The everyday mannerisms that characterise a person – their rhythms of speech, bodily comportment and taste preferences – are testament to such influences, while also highlighting the complex interplay between inheritance and environment. We can also observe that some events and relational connections have enduring impacts upon the self, such as the resonance of educational opportunity or perhaps the grief of a personal loss. Others touch us only fleetingly, however, and are quickly absorbed in the passing flow of life, seemingly forgotten (Conradson & McKay 2007: 167–168).

Thereby selfhood becomes 'always a hybrid achievement, emerging out of a diverse range of connections' (ibid.: 168). As such, Conradson and McKay (2007) are explicit in theorising the relationality of migrants' subjectivities and I concur with their view that it is in through the transformation of people's connections to others that international migration might reconfigure subjectivities:

> Geographical mobility inevitably changes the relations we have with emplaced configurations of people and events, while at the same time bringing us into contact with new and different ecologies of place (Conradson 2005). For transnational migrants, the relational effects of mobility may be particularly significant. A person might choose, for example, to exchange a sense of community and reciprocity in a village setting for the economic opportunity yet relative anonymity of a major city. In the process, their selves will be shaped by new relations in the destination setting, as well as through the distance obtained from those that characterise the sending context. Mobility thus provides opportunities for new forms of subjectivity and emotion to emerge, whether broadly positive or negative (Sheller & Urry 2006) (Conradson & McKay 2007: 168).

In order to extend existing research on migrants' subjectivities, in this book I build on the broader literatures on migration and transnationalism that I have mapped above. My usage of the term 'intimacy' departs, however, from that of Pratt and Rosner (2012) who adopt it in bringing together authors focusing on bodies, households, emotions, affects, and materialities. Furthermore, in work invoking the terminology of intimate subjectivities, the focus has been primarily on the negotiation of work and either motherhood or marriage (e.g. Constable 2016; Mahdavi 2016). In contrast, sociological studies of personal life have at their heart the notion of an 'intimacy of the self' (Jamieson 1998) such that intimacy, in this book, is not used as a synonym for physical closeness, proximity, or sexual relationships. Instead, rather than sexual or couple relations, my focus will be inclusive of a broader range of personal relationships among communities, friends, and families as well. Chapter 2 further explores related literatures and my approach to intimacy, but first I outline the structure of the book.

Approaching Intimacy: The Structure of the Book

First, in Chapter 2, I review the extensive interdisciplinary literatures on intimacy and related concepts, setting out in much more detail the theoretical approach this book takes and why. Most notably, the second half of the chapter works to establish how the concept of 'intimate subjectivities' focuses attention on the spatialities of intimacy. I note that bodies remain important in understanding and analysing intimacy, since our experience of personal relationships is corporeal and emotional, and our enactment of them embodied and performative. For Jamieson (1998), one of the tasks is to evaluate whether intimacy exists in people's personal lives. In contrast, I am concerned with exploring how the desire (or not) for something we might understand as 'intimacy' informs practices of everyday relational life and the way in which people talk about and negotiate their personal relationships. While Jamieson (1998: 8) recognises that 'intimacy' often privileges emotional disclosure at the expense of recognising 'silent' practices of loving, caring, affection, and sharing, she also stresses that close association and detailed cognitive knowledge of another are insufficient to ensure intimacy, since a degree of sympathy and empathy are key.

Chapter 3 then outlines the Gulf Cooperation Council (GCC) context, especially the economic transformation of Dubai, as well the methods for the study of British migrants, who they are, and includes a reflexive statement. One of the important contributions of Chapter 3 to understanding intimacy in the lives of British migrants is that it examines the *Kafala* sponsorship system which, as mentioned briefly earlier, determines their status in Dubai as non-citizens whose residence is temporary. While in practice many Britons live much longer in Dubai than they perhaps anticipated (the length of residence among interviewees ranged between one and twenty-five years), this legislative environment nevertheless frames their settlement and the cultures of intimacy that emerge. Another significant aspect of everyday life in Dubai that is introduced and explored in Chapter 3 regards the racialised and classed social hierarchy. Again, the implications for British migrants' intimate subjectivities can be traced throughout the empirical chapters.

Chapters 4–7 are the empirical chapters analysing, in turn, the partial separation of British migrants within the wider socio-spatial segregation of the city; the establishment of community organisations and friendship; sex, desire and romance; and family life. In more detail, the objectives of the empirical chapters are as follows. The first empirical chapter discusses the making of British 'expatriate' subjectivities in Dubai. I explore British migrants' imaginaries of Dubai and Emirati culture, as well as their encounters with the national legislation and bureaucracy. It is important for later chapters to understand how British 'expatriate' subjectivities come to be marked as distinctive, as well as how racialised ontologies of difference help regulate the production of intimate subjectivities in everyday life in Dubai. British migrants rarely develop meaningful relationships with Emirati nationals beyond workplace collegiality, as a result of complex processes of national boundary making arising from both communities. Therefore, explicating how British migrants relate to citizens and other migrants as I do in Chapter 4 helps to explain the dominant focus of subsequent empirical chapters on relationships between British nationals or between Britons and a larger group of mainly white, Western migrants understood as 'expatriates'. I suggest that the 'expatriate' lifestyle is distinctly spatialised in Dubai, taking place largely within privatised leisure spaces with particular implications for the regulation of intimacy. Such environments serve to exclude low-income migrant workers, limiting contact to client and work-based relationships. I do explore how some of these encounters play out, especially those taking place in what might be called 'intimate' spaces (the taxi, home, or nightclub, for example), but I argue that the establishment of British 'expatriate' subjectivities for the most part precludes intimacy emerging in these relations. Significantly, Jamieson (1998: 1) suggests that intimacy requires equality in the relationship in which it is to be enacted: 'intimacy across genders, generations, classes and races can only take on this character if the participants can remove social barriers and transcend structural inequalities'. Arguably, the politicisation of intimacy acquires further significance and new meanings in the transnational spaces of Dubai which are characterised by diversity, segregation and shifting hierarchies. As such, this politicisation is key to my enquiry also.

Chapter 5 builds on the preceding discussion to examine more deeply the social lives of British migrants. I first explore the various types of voluntary community organisations (VCOs) that exist in Dubai and cater for British migrants' expressed desire to feel a sense of community and meet new people who are similar to them after relocating. Some of these clubs are highly exclusive, based on the nationality of their membership, while others are more implicitly racialised and classed spaces. The significance of intimacy in grounding migrants in their new city becomes apparent in this chapter, with experiences of old hobbies and new activities leading to the development of friendships among the participants. As such, in the second half of Chapter 5, I go on to explore the significance of friendship itself, identifying some of the transformations in friendship practices among British migrant residents as they negotiate settlement, however temporarily, in Dubai.

From the discussion of VCOs and friendship in Chapter 5, the distinctive lifestyles of single and married British migrants begin to emerge. Chapters 6 and 7 explore the intimate subjectivities of these two (internally heterogeneous) groups in turn. Chapter 6 focuses almost exclusively on single Britons, examining the night life and associated cultures of sex, desire, and romance. In the chapter, I show how straight British migrants enact new intimate subjectivities in Dubai, contesting not only the heteronormative couple relations of their peers in the UK and the married British migrants they encounter after relocation, but also contravening the UAEs Decency Laws. Geographies of displacement and the amplified social status of British migrants in Dubai appear to encourage gendered practices of sexuality and friendship to be performed through the night-time spaces of the globalising city.

The final empirical chapter focuses on family life, but maintains a critical approach to the work involved in constructing a heteronormative domestic home life. The chapter highlights how these British families, like those of other migrants (e.g. see Pratt 2012 on migrant domestic workers) are embedded in the broader political economy of global work. In the case of Britons in Dubai, this demands of them gendered relationships to work and home which can challenge the sense of equality experienced by married women. The chapter discusses the notion of 'family time', domestic homemaking as a family practice, the continued significance of relationships with adult children, and familial relationships with parents and siblings 'left behind', focusing on the way in which migration to Dubai necessitates the renegotiation of family ties and practices.

Overall, then, the book examines different kinds of personal relationships in distinct chapters, organised as we might anticipate from both existing academic theory and everyday life. However, having explored British migrants' intimate subjectivities ethnographically, I am also able to draw attention to the moments when in practice, or discursively, different kinds of relationship inform each other. For example, in Chapter 5, interviewees describe how their friendships are informed by their couple relations, more specifically gendered parenting and working practices. The final chapter offers some concluding remarks on intimacy, focused especially on the spatialities of globalisation.

Notes

1   This figure comes from the first of two reports compiled by the Institute of Public Policy Research – *Brits Abroad* (Sriskandarajah & Drew 2006) and *Global Brit* (Finch 2010) – that provide the best statistical estimates currently available on the numbers and destinations of international British migrants. The authors of these studies readily admit that the IPPR figures are compromised by the difficulties in capturing this kind of quantitative data (e.g. Sriskandarajah & Drew 2006: 8). To get the best possible estimate they bring together several empirical sources, each incomplete in its coverage and with varying degrees of reliability. The International Passenger Survey, for example, is widely acknowledged to be of limited value since the sample is so small. The IPPR estimates of the numbers of Britons in the UAE, however, reveals some more specific limitations of this methodology for this particular geographical site. A more widely applicable problem with census data – that it is produced only once a decade and quickly becomes out of date – is magnified in the UAE where extremely rapid population increases were seen after the millennium, such that uprating the figures based on previous growth patterns still leaves the figures underestimating the likely numbers. The Department of Work and Pensions data on people receiving their pensions is also virtually irrelevant in the UAE because the British nationals there need to be of working age or accompanying their employed spouse/parent. The UAE was also not considered a high-risk country so the Britons I spoke with were very rarely registering with the British consulate. Finally, the Labour Force Survey captures returnees across the UK so might be more reliable for well-established expatriate destinations, but would not have registered the explosion in out-migration to Dubai until some years later. The actual number of Britons resident in the UAE during the period of this ethnographic study was, therefore, likely to be significantly higher than these figures would suggest and was estimated to be 100,000 by interviewees.
2   In exploring gendered identities, my understanding also draws on a large body of work in geography and beyond that approaches femininities and masculinities as plural, diverse and spatially contingent (e.g. Berg & Longhurst 2003; van Hoven & Hörschelmann 2005). Hopkins and Noble (2009: 815) reflect on a disciplinary move to provide more nuanced accounts of masculinities that embrace such spatial contingencies.
3   Pratt and Rosner's (2012) appeal to consider the 'specificity of processes in distinctive places' through a dialectical notion of the intimate–global does, however, resonate with other critical projects not framed by feminism e.g. that of 'global ethnography' (Burawoy et al. 2000).
4   For some, this is framed in the language of affect, for others emotion, and some scholars incorporate both. For example, Kevin Dunn notes: 'Transnationalism is about encounters between different bodies which leads to all kinds of intimacies and emotions, some that generate sharing and exchange and others which lead to tension, friction and even hostility and anger. And importantly these intimacies are visceral encounters; they trigger embodied and affective responses' (Dunn 2009: 6). As I consider further in Chapter 2, affect and emotion are very different theoretical terms in the field of geography, but my own approach, informed by my methodological choices, highlights the emotional geographies of migrants' intimate subjectivities.

5   As early as 2002, Russell King wrote of 'love migrations' as 'an essential component of the "new map of migration" in Europe. King (2002) drew our attention to the way in which contemporary migrations were increasingly being explained with reference to individual and personal factors. While highlighting the 'libidinal factor' for students and tourists, he suggested that 'love migration can probably be found in all types of migration' and pointed to the transformation in technologies 'shrinking Europe' as increasing the chances of 'transnational intimacy' and 'transnational love' being maintained. In the European context, King understood major cities as being the principal nodes for an 'intensification of cross-national personal contact, relationships, partnerships and marriages', for here several important sociological factors – namely, the expansion of linguistic competence and 'the expansion of "global experience" industries (tourism, travel, leisure, education, networking)' – coalesced to 'produce an expansion of individual transnational interfaces' (99–100).

Chapter Two
Geographies of Intimacy

In 2008, Gill Valentine's call to develop a new geography of intimacy revealed, in the associated review of existing publications, a discipline unwittingly neglectful of the significance of 'the ties that bind', especially families. Valentine (2008) suggested that, in spite of extensive critical attention to the related fields of sexuality, childhood/youth, and parenting, geographers were seemingly shy to directly engage with intimate relations, perhaps as a result of their affective register. Geographers' concerns with these three thematics had emerged as relatively discrete sub-disciplinary areas, even while the empirical work they generated, Valentine argued (2008: 2097), can be seen to share not only 'a focus on bodies and identities' but also a desire 'to challenge traditional public-private dichotomies' and 'to understand the more complex inter-spatiality of personal lives'. This neglect was in distinct contrast to a burgeoning sociological scholarship focusing directly on families and intimacy (e.g. Beck & Beck-Gernsheim 2005; Jamieson 1998; Morgan 1996; Smart 2007). Fortunately, in the decade following, it is possible to witness a new confidence in geographical scholarship engaging with intimate life. Prominent review papers and editorials on friendship (Bunnell et al. 2012), love (Morrison et al. 2012), families (Harker & Martin 2012) and intimacy (Moss & Donovan 2017; Oswin & Olund 2010; Pain & Staeheli 2014) are suggestive of a discipline rethinking the significance of intimacy in the spatialisation of our practices, subjectivities and everyday lives. This chapter explores these recent geographical literatures as a resource through which to examine the particularities of my own empirical research in later chapters. In addition, it

Transnational Geographies of the Heart: Intimate Subjectivities in a Globalising City, First Edition. Katie Walsh.
© 2018 John Wiley & Sons Ltd. Published 2018 by John Wiley & Sons Ltd.

examines wider interdisciplinary literatures on intimacy, friendship and family to explore what they might offer geographers interested in intimacy. I bring these relatively distinct debates together in order to establish some of the features we might push for in a geographical framework and outline my approach to analysing the intersections of intimacy and transnationalism in everyday life through the subsequent empirical chapters.

My own analysis of British migrants' intimate lives presented in this book has benefited enormously from this explosion in disciplinary attention on intimacy. In writing my doctoral thesis in 2004, I struggled to find appropriate geographical sources and was forced instead to navigate sociological theories of individualisation, with their often gloomy takes on contemporary life and abstract conceptualisation of relationships (Bauman 2003; Beck & Beck-Gernsheim 1995; Giddens 1992). Writing this chapter a decade later, I am inspired not only by this emerging literature in geography but also by an increasingly theoretical interdisciplinary engagement with family, emotion and sexuality among scholars working on migration specifically (e.g. Baldassar & Merla 2014; Mai & King 2009; Skrbiš 2008), as discussed in Chapter 1. I would argue, however, that there is much more to understand. As such, in this chapter, I will demonstrate that attention to the nexus of intimacy and migration will be critical to our developing understanding of the everyday affective geographies of belonging and home in a world of increasing movements of people. I suggest that we must pay critical attention to the intersections of race, ethnicity, gender, class, religion, nationality and migration status in analysing intimate subjectivities as embodied and emplaced in particular sites. The negotiation of personal relationships is shaped not only by our social locations, but also by the way in which these relationships are mobilised spatially through cultures of intimacy that shape the norms and practices of our social lives.

Families and 'Family Life'

On the basis of her aforementioned extensive review, Gill Valentine (2008: 2101) suggested that studies of families are a 'peculiar absent presence' within geography. She noted that relationships between parents and children/young people had been a concern of the geographies of parenting, but they had rarely been understood in terms of family life and familial relations in a broader sense. Furthermore, Valentine (2008) suggested that a sub-disciplinary literature on geographies of sexuality, by then well established, had paid little attention to the personal relationships through which intimacy is constituted, attributing this to the way in which feminist and queer approaches had partly been mobilised through a critique of 'the family' as a heteronormative construction. As Harker and Martin later elaborated in their editorial to a Special Issue on familial relations,

While geographers have provided rich analyses of gender, social reproduction, and capitalism; sexuality and space; and the boundaries of public and private space, these approaches have circled around 'the family'. There are good reasons for this. Scholars engaged with these themes have sought analytic starting points that denaturalize the biological family as a site of feminine labour, normative sexuality, and depoliticization (Harker & Martin 2012: 768–769).

Empirical contributions have since attempted to explore what the spatialities of family life might look like (Pratt 2012; Valentine & Hughes 2012). Yet, this absent presence of family in geographical scholarship still persists to a degree and stands in stark contrast to what we find in sociological literatures (Valentine 2008), where family life was being examined with renewed vigour from the early 1990s (e.g. Finch & Mason 1993; Morgan 1996).

Indeed, mainstream sociological theorists of late-modernity (e.g. Beck & Beck-Gernsheim 1995; Giddens 1992) were highlighting intimacy as an important site of enquiry, challenging the low-status evaluation more usually bestowed on work on the 'trivialities' of family life (Smart & Neale 1993). At the same time, empirically based studies by sociologists were revealing changes in family life and relationships, with higher rates of divorce, increasing cohabitation and 'reconstituted families' contributing to an increased diversity in contemporary family forms (Smart & Neale 1999). The reification of family and kinship as a 'natural' form of belonging was also being thoroughly challenged: new forms of relatedness (in particular, cohabitation, divorce, step-families, gay and lesbian families, adoption) and new reproductive technologies questioned kinship as 'natural' or 'given' by exposing the self-conscious project of kinship. Weston (1991) had coined the term 'families of choice' to describe the challenge to, and reproduction of, conventional forms of family being generated by same-sex partners. Janet Finch's work (Finch and Mason 1993) on kinship especially, revealed how these kinds of wider social changes were both responsible for, and helped to shape, a transformation in notions of moral obligation between kin more generally. Rather than being solely dutiful, Finch (1989) saw obligations and commitments of kinship as also arising in part from empathy and affinity. Morgan (1996) suggests that broader feminist contributions on care also started to consider women's agency, thereby complicating the theorisation of home and patriarchal gender relations in ways that allowed familial relationships to be understood differently. With these multiple threads transforming sociological study, Carol Smart and Bren Neale's (1999) study on post-divorce family life summarised the position of the discipline as it came to a new understanding in the late 1990s:

> Clearly we cannot use the term 'the family' without the inverted commas any more. It is far too naive a concept. It implies that there is a naturalistic grouping which always, everywhere, is the family with its fixed gender roles. It distorts differences of

class, race, region and so on. Because this traditional definition is based on a unit of reproduction, it presumes heterosexuality as the norm, and it presumes that only those linked by blood or marriage can be part of a family (Smart & Neale 1999: 20).

'The family', then, was increasingly redundant in sociological theory. Rather than a static and homogenous institutional understanding, David Morgan (1996) argued, we needed to shift sociological analysis to attend to the 'doing' of family.

Later revisiting this idea, Morgan (2013) argued that 'family' should be seen not as a noun to describe particular or static models of social life, but rather as an adjective – as in family life, family processes, family events, and family practices – or even a verb, to keep a sense of 'doing family' as a set of social activities. Furthermore, he suggested that the language of practice is productive, since it highlights not only this sense of the active dimension of families, but also a sense of the everyday life-events and unremarkable routines through which family is constituted and a sense of fluidity, since who 'counts' as family is somewhat flexible. Responding to criticism, Morgan (2013: 7) argued strongly in this later account that family practices are informed by context, 'structuration as a set of processes rather than fixed external structures', such that individuals 'come into (through marriage or parenthood, say) a set of practices that are already partially shaped by legal prescriptions, economic constraints and cultural definitions'. He notes that family practices are usually taken -for-granted, tacit parts of social life, but that the constitutive character of family practices is revealed when routines, expectations and assumptions about family are challenged or become problematic for some reason. I would suggest that migration is precisely such an event in which family comes into question. Therefore, the reflexivity that British migrants show towards their relationships, including family practices, should perhaps not be a surprise as they navigate changing sets of norms in relocating to Dubai, especially shifts in gendered practices of work and home (see Chapter 7; I also discuss Morgan's take on 'family practices' further in the final section of this chapter).

For Finch (2007), there is an additional dimension in the constitution of families, that is 'displaying family', a notion through which she aims to 'emphasise the fundamentally social nature of family practices, where the meaning of one's actions has to be both conveyed to and understood by relevant others if those actions are to be effective as constituting "family" practices' (ibid.: 66). Drawing on the links made between interpersonal intimacy and identity (Beck 1992; Beck & Beck-Gernsheim 1995; Giddens 1992), Finch suggests:

> In a world where families are defined by the qualitative character of the relationships rather than by membership, and where individual identities are deeply bound up with those relationships, all relationships require an element of display to sustain them as family relationships (Finch 2007: 71).

As such, she argues that the need to display family has intensified for contemporary families and is not restricted to non-conventional family forms.

But what of 'families' in migration studies? Transnational families – defined as those where the members 'live some or most of the time separated from each other, yet hold together and create something that can be seen as a feeling of collective welfare and unity, namely "family-hood", even across national borders' (Bryceson & Vuorela 2002: 3) – have been studied primarily through ideas about care and/or motherhood. For Loretta Baldassar and Laura Merla (2014: 7), care is the dominant theoretical tool for analysing transnational families, and they make a case for caregiving as a key practice of 'doing family' (drawing on Morgan 1996), even while separation, and a consequent lack of co-presence, dominates the everyday lives of transnational family members. This transnational caregiving, they argue, 'binds members together in intergenerational networks of reciprocity and obligation, love and trust, that are simultaneously fraught with tension, contest and relations of unequal power' (ibid.: 7). Family life is characterised, Merla and Baldassar (2014) suggest, by the negotiation, monitoring and management of care exchanges and commitments across the life course and, therefore, should be understood as circulating among family members over time and distance.

For other scholars, especially those motivated by advocacy with migrant domestic workers, there is a need to focus attention on the violent reorganisation of intimacy and family life emerging from global labour markets and temporary migration regimes, especially the impact this has on relationships between mothers and their children. In *Global Woman*, Barbara Ehrenreich and Arlie Hochschild (2003) argued that the feminisation of labour migration was resulting in a care deficit in sending countries like the Philippines, where women were leaving their children with fathers and extended kin in order to work overseas and earn the money for food, healthcare and education. Geographers have been among those who have critiqued Ehrenreich and Hochschild's 'global care chains' thesis for its victim narrative and have argued that we instead need to theorise migrant women as self-reflexive agents with diverse responses to global migration (e.g. Silvey & Lawson 1999). Rhacel Parreñas' (2001, 2005) extensive research with migrant children and their mothers reveals some of the complexities of these debates. Parreñas (2001) demonstrates the efforts of migrant mothers to maintain intimacy with their children in an attempt to counter the negative stereotypes of maternal abandonment that shape discussion of such women. However, she also (2005) reveals that children in the Philippines, whose mothers may be absent for their entire childhood, often fail to recognise their mother's love, due to the strong persistence of gendered parenting roles that equate mothering with a *proximate* nurturing role. In contrast, in *Families Apart* Pratt (2012) acknowledges the theoretical need to challenge gendered and ethnocentric norms of love and intimacy, but states as her aim the exposition of the trauma of separation and the destructiveness of distance. Among the women using the Philippine Women Centre, Vancouver, with whom Pratt collaborated, it was common for visits home to be short and infrequent, typically two weeks every three years (Pratt 2012: 48). As a result, mothers and children spoke of their non-recognition of each other and Pratt (2012) highlights their ongoing experiences of dislocation even when they are reunited physically in one place.

In contrast to the transnational families who feature in these existing studies, most British lead-migrants to Dubai are of an age where parental care is not yet a concern and/or their income level allows them to sponsor the visas of their spouse and children to accompany them, so that it is generally only adult children who are 'left behind' or who leave Dubai for higher education and their own careers (see Gardner 2011 for a comparison of low-income migrants in GCC countries). As such, frameworks focusing on international circulations of care are less relevant to my analysis (although see Chapter 7). Instead, the dominant concerns of British migrants resonate more with a narrower literature on the families of transnational professionals, which has provided insight, especially, into the experiences of the 'accompanying' female spouse (see Coles & Fechter 2008; Hardill 1998; Hardill & MacDonald 1998). Indeed, rather than share the transnational experiences of mothering depicted by existing interdisciplinary scholarship, British migrant women are instead able to spend more time in proximity with young children than they might in the UK where dual incomes are more often necessary, As such, they often value the opportunity that migration offers them to engage in quality parenting time by displacing the more messy and repetitive tasks onto migrant domestic workers that they can employ in their homes as a result of an increased disposable income (see Chapters 4 and 7). While a few parents still choose to send their children to boarding school in the UK, this is increasingly rare and usually narrated as the child's decision in response to the family's constant mobility. Therefore, the Britons I spoke with may have had adult children or parents, siblings and cousins living elsewhere, but the negotiation of these relationships did not emerge from my ethnography as especially concerning (as it might have if separation was from younger children and husbands). Exceptions were evident in the lead-up to Christmas where return visits were anticipated and in cases where parents were becoming elderly. I explore this all further in Chapter 7, but there is clearly a lot more that could be understood from a more intentional and sustained research focus on the transnational relations of privileged British migrants and how their transnational families are sustained, felt and practised in everyday life.

A critical approach to family also recognises the moments when families function to (re)produce inequalities or for some other reason do not feel ideal in terms of their intimacy. Chapter 7, for instance, highlights instances of the difficulties negotiating transnational family ties with siblings and parents. It also documents the shifting intimacy between some married couples as they navigate the gendered expectations and identities at play in transnational spaces. Therefore, it is important not to privilege and celebrate the intimacy of heteronormative family life as somehow more straightforward and automatically more significant than other kinds of intimate relationship. Roseneil and Budgeon (2004: 129) remind us that it cannot be assumed that family relations are sites for the practice and experience of love and, while the concept of family remains a cultural ideal that is central to our social organisation and the construction of our social and personal identities in contemporary Britain and more widely, 'much that matters to people in terms of intimacy and care increasingly takes place beyond the "family", between partners who are not living together "as family", and within networks of friends.' So, friendship is becoming increasingly significant in contemporary cultures of intimacy, perhaps even as 'families of choice' (Roseneil & Budgeon 2004): more and more people are living periods of their life outside marriage and as a result are sharing their domestic and emotional life with friends instead (Heath 2004; Pahl & Spencer 2004; Roseneil & Budgeon 2004). I therefore now turn to consider the geographies of friendship.

Geographies of Friendship

With Tim Bunnell et al.'s (2012) recent interdisciplinary review, friendship is at last firmly on the agenda in geographical scholarship. Friendship was given a high profile when Nigel Thrift (2005: 146) suggested that friendship has a role to play in 'keeping cities resilient and caring', but his theoretical portrayal of friendship as a 'light-touch' model of intimacy (and as distinct from, and less significant than, loving couple relationships) is typical of the way in which friendship was, until recently, rarely examined deeply by geographers (the notable exception being Valentine 1993). In contrast, the significance of friendship was recognised much earlier, and examined more fully, in cognate disciplines, especially anthropology (e.g. Bell & Coleman 1999) and sociology (Allan 1996; Eve 2002; Fehr 2000; Heath 2004; Pahl 2000, 2002; Roseneil and Budgeon 2004; Spencer & Pahl 2006). Indeed, sociologists have long identified friendship as an important source of love and meaning (Fehr 2000.) As Pahl (2000: 1) argues, a 'special friend can share one's deepest thoughts, hopes and fears and provide "another self" to share the vicissitudes of life.' Likewise, in his sociological analysis of friendship, Allan (1996) suggests that the intimacy of friendship is thought to be enormously important in helping people to cope with the demands of daily living within a wider society. While 'families' have so much moral and symbolic significance in our imagined geographies of intimacy, friends it seems are quietly important. For Allan (1996: 110), friends provide much of the routine interaction of sociability in shared activities, discussion, and confirming each other's identities, as well as providing mutual practical and emotional support. Like family relations, friendships are understood in this work as made through 'active, ongoing and necessarily reciprocal work' (Bunnell et al. 2012, drawing on Vertovec).

Bunnell et al. (2012: 491) suggest that geography 'is important in the making, maintenance and dissolution of friendships, as well as in the types of friendships that are important within particular time-settings'. Sophie Bowlby's (2011) work on friendship and care demonstrates well some of these spatialities of friendship. Firstly, she notes that the larger public setting is important since state welfare provision will inform the expectations of informal care between family and friends. Secondly, she identifies that our social locations – in terms of class, ethnicity, gender – will inform with whom we seek to be friends (on sexuality see Valentine 1993). Bunnell et al. (2012: 491) echo this concern, arguing that it is important for geographers to recognise that 'friendship is not merely important in its own right but also plays a role in broader processes of social ordering and transformation', such that class, gender and sexuality might be '(re)produced and strengthened through the work of friendship.' Thirdly, Bowlby (2011: 614) identifies the role of the domestic home space as being important, yet under-examined, as 'a place in which caring between co-residents is a social expectation'. Indeed, the very meaning of 'home' is wrapped up with this idea of care, but this should not be equated with the family. Bowlby reminds us that we know very little about other forms of households, such as those among friends. These three spatialities of friendship are evident across the chapters in this book. The social ordering of friendship is certainly supported by the evidence in the friendships of British migrants in Dubai, with race, class, nationality and gender especially significant. Chapter 4 demonstrates how urban encounters in public and privatised spaces in Dubai (e.g. leisure spaces, the home, or the beach) are shaped by structural hierarchies that work to 'shut down' opportunities for meaningful social interaction across difference in everyday life. Chapter 5 then reveals how this racialised and classed segregation plays out in the emergence of British and 'expatriate' institutions that further shape how friendships develop in these restricted 'social contexts' (Bowlby 2011). I also reflect on the importance of various spaces in the geographies of friendship, including the home, but (since my focus is not on care) also spaces of leisure. Finally, I suggest that international relocation to Dubai, and consequent navigation of the spatialities of intimacy across notions of being at home and away, could transform the meaning of friendship itself. In terms of efforts to understand migrants' everyday lives more widely, friendship has mostly been obscured by the concepts of social capital, networks and community (but see Conradson & Latham 2005b on Antipodean transmigrants in London).

The challenge for geographical enquiry, as articulated by Bunnell et al. (2012: 492), is to integrate 'understandings of friendship that are grounded in sites and everyday spatial practices while also being attentive to the ways in which intimate relationships are embedded in wider social and political formations.' The coproduction of intimacy is something already widely recognised in studies of sexuality and heteronormativity, so I elucidate this further in the next section.

Couple Ties: Heteronormativity, Governance, and Love

The attention of geographers towards couple ties was at first indirect, through their interrogation of sexuality. The publication of Bell and Valentine's (1995) book *Mapping Desire* was followed by the emergence of an identifiable sub-discipline of sexuality studies within geography (e.g. Browne et al. 2007). Focusing directly on heterosexuality has rarely been part of that critical project, but recently queer theory has been instrumental in exposing heteronormativity as a set of norms that privilege marriage, family and biological reproduction in the organisation and expression of heterosexuality (Manalansan 2006). As Corber and Valocchi (2006: 4) remind us:

> Although they overlap, heteronormativity and heterosexuality are not co-extensive and cannot be reduced to each other…there may be modes of organising sexual relations between straights that are not heteronormative, just as there may be modes of organizing sexual relations between gays and between lesbians that are.

Hubbard's (2008) review paper on the geographies of heteronormativity serves both as an empirical reminder and theoretical assertion 'to explore the many possible articulations of heterosexual desire that are included or excluded within a dominant construction of heteronormality.' This theoretical assertion emerged from his own research on sex work that considers the regulation of heterosexualities seen as excessive, perverse or immoral (Hubbard 2008).

The governance of intimacy was also the focus of a Special Issue of *Society and Space*, edited by Natalie Oswin and Eric Olund (2010), in which they convincingly argued that '"the intimate" is not only the sphere of individual subjectification' but, rather, 'Kinship, procreation, cohabitation, family, sexual relations, love – indeed all forms of close affective encounter – are as much matters of state as they are matters of the heart.' To make their case, Oswin and Olund (2010) draw on scholars working at the intersection of feminist, queer and postcolonial studies, who have as their starting point an understanding of the intimate as a co-production with the public. They identify the work of Lauren Berlant (2000) and Elizabeth Povinelli (2006) as especially useful since their work has helped forge an understanding of intimacy in which scale is unfixed, such that 'intimate relations cannot be considered synonymous with the body or the household, locations which then simply mirror larger social relations through their capacity to oppress or liberate at closer physical proximity' (Oswin & Olund 2010: 60)[1]. It is worth, therefore, considering the contributions of these scholars here in more depth. For Berlant (2000: 2) it is our attachments to others that make people public, 'producing transpersonal identities and subjectivities', and thereby linking individual lives to a collective sense of society. Yet she notes that normative ideologies promote certain expressions of relationship ('love, community, patriotism'), while discrediting others; 'desires for intimacy that bypass the couple or the life narrative it generates have no alternative plots, let alone few laws and stable spaces of culture in which to clarify and to cultivate them' (Berlant 2000: 5). Berlant (2000: 3) reminds us, therefore, that while scholars may consider categories of public and private to be outmoded, there is a continuing attachment to this division since 'the discourse world described by the public and the private has, historically, organized and justified other legally and conventionally based forms of social division.' She gives the examples of 'male and female, work and family, colonizer and colonized, friend and love, hetero and home, "unmarked" personhood versus racial-, ethnic-, and class-marked identities' as being mapped across the public/private boundary and suggests that the allure of this boundary lies in the way it 'can reverberate and make the world intelligible' (ibid.: 3). These are themes that strongly resonate with the concerns of Elizabeth Povinelli (2006) in *The Empire of Love*. This work, a theoretical reflection on intimacy in the contrasting, yet related, social worlds of an indigenous community in Belyuen, in the Northern Territory of Australia, and among radical queers in the US, reminds us of the way in which intimacy is culturally coded, apprehended differently by people living different lives. The recognition of intimacy norms as contingent on space and time is vital for an understanding of migrants' intimate subjectivities, since migrants must navigate both the cultural norms and habits they carry with them as embodied (though reflexive) knowledge, as well as the collective discourses and legislative rules of their place of residence.

An emerging body of work in the early 2000s took a queer theoretical perspective to analyse flows of gay and lesbian migrants more specifically (e.g. Binnie 2004; Gorman-Murray 2007; Knopp 2004) and pay critical attention to the relationship between sexuality and citizenship (Luibheid 2005; Puar 2006). The work of Manalansan (2006: 2) is exceptional, however, in adopting a queer theoretical perspective to consider the functioning of heteronormativity in relation to heterosexual migrants 'as a way to complicate and re-examine assumptions and concepts that unwittingly reify normative notions of gender and sexuality.' Manalansan (2006) suggests that we need critical insight into the way in which heteronormativity has implicitly structured scholarly analysis of migrant subjectivities and practices, as well as impacting upon migrants' lives. For example, with respect to Filipino migrants, he argues that the application of queer theory promises a way of 'unsettling normative conceptions of parenthood, maternal love, and care by not locking them into specific gendered or married bodies' and by repositioning migrant domestic workers as 'viable desiring subjects.' A collection of ethnographic papers on heterosexuality and migration that I had the privilege of co-editing also set out to focus explicitly on some of the diversity evident in situated performances of migrant heterosexualities (see Walsh, Shen & Willis 2008). When we begin to consider heterosexualities as specific and contingent, as emerging in particular local spaces infused by transnational flows, it becomes clear that a range of heterosexualities are enacted. It also becomes apparent that individual and collective articulations of these identities may shift as migrants move through and constitute transnational social fields, negotiating normative cultures in multiple sites. As a result, the analysis in Chapters 6 and 7 reveals the considerable everyday work that goes into the reproduction of heterosexualities and heteronormative relationships more frequently taken for granted in the literature. Therefore, while I cannot claim to present a similarly radical project as that encouraged by queer theorists, informed by these perspectives I do attempt to reveal both the gendered work and emotional labour (Hochschild 1996) that is entailed in the production of heteronormative family life (Chapter 7), as well as the resistance to these constructions by single British migrants as they find themselves negotiating their place in Dubai and state attempts to regulate the intimate lives of migrants working in the UAE (Chapter 6).

However, a turn to thinking about couple intimacy in geography can be understood not solely within this sexualities framework (and more recently queer framework), but also as part of a wider disciplinary shift towards the recognition of emotions in our everyday lives and subjectivities. Carey-Ann Morrison, Lynda Johnston, and Robyn Longhurst (2012: 506) recently called to geographers to engage more explicitly and critically with love, especially love's spatiality, relationality and politics. In terms of the spatiality of love, they argue that both material places and discursive spaces are important in shaping feelings and practices of love. Sedgwick's (1999) *A Dialogue on Love* provides the impetus for their understanding of love as relational, by which they mean 'thinking through the relations and spaces between and among individuals, groups and objects' (Morrison et al. 2012: 513). They draw on Lawson, and Sedgwick, in thinking about the circulation of love in multiple directions and the complexity of power in which love offers hope to go beyond our theoretical reductionism. Finally, drawing on Ahmed (2004) and Berlant, like those theorising sexuality (see above), Morrison, Johnston and Longhurst remind us of the impact of heteronormativity in determining which acts of love are acceptable, should be celebrated, and are enshrined in notions of respectability in terms of both gender and citizenship: 'love being normatively mapped on to the spaces and subjectivities of heterosexuality alert us to the fact that love is always political,' such that the productive line of enquiry is to 'unpack the ways in which power circulates in love relationships' (Morrison et al. 2012: 515).

Love becomes apparent in the production of intimate subjectivities in this book, not only in relation to couples but also in families and friendships. Nevertheless, the dominant idea of the couple relationship as being the most intimate in our lives is a persistent cultural script. In the next section I look at the sociological literatures which tackle the language of intimacy more directly in their theorisation of interpersonal relationships.

Intimacy and the Sociological Study of Personal Life

So far in this chapter I have engaged with rather distinct sets of geographical and interdisciplinary literatures, especially those concerned with families, friendship, love and heterosexuality. Reflecting on the breadth of this existing knowledge reveals that there are a large number of conceptual tools through which to examine the personal relationships of the British migrants I met in Dubai. These tools will certainly not be abandoned in the chapters that follow, yet here I add another conceptual term: intimacy. Why is this necessary and what exactly do I mean by intimacy? Though, like many other geographers, my interest in intimacy has been prompted by studies of identity and sexuality, and informed more recently by queer theory, in theorising intimacy my starting point was instead what we might term a 'sociology of personal life'. In this section, I examine this literature in more detail before proceeding to outline my understanding of and approach to intimacy in this book.

Just after the millennium, Jamieson (1998: 1) reflected that 'intimacy' had already become 'a fashionable word both in the social sciences and in popular self-help books advising on the art of good relationships' and that it was often used to refer to 'a very specific sort of knowing, loving and "being close" to another person.' The 'intimate relationship', Jamieson argued (1998: 2) had emerged as 'an idealized version of personal life' in its contribution to our emotional well-being. At the heart of 'meaningful social life', she suggested, we now have a notion of intimacy in which the emphasis is on 'mutual disclosure', highlighting practices of talking and listening such that feelings and thoughts can be shared (ibid.: 2). Jamieson (1998: 1) terms this an 'intimacy of the self rather than an intimacy of the body, although the completeness of intimacy of the self may be enhanced by bodily intimacy.' 'Intimacy' is not, therefore, used in this book as a synonym for physical closeness, or sexual relationships, and the focus will instead be on a broader range of personal relationships among communities, friends and families, not just couples. I note, however, that bodies remain important in understanding and analysing intimacy, since our experience of personal relationships is corporeal and emotional, and our enactment of them embodied and performative (more on this below). For Jamieson (1998), one of the tasks is to evaluate whether this kind of intimacy exists in people's personal lives through various kinds of interpersonal relationships. In contrast, I am concerned with exploring how the desire (or not) for something we might understand as 'intimacy' informs practices of everyday relational life and the way in which people talk about and negotiate their personal relationships in contexts of international migration, especially in the globalising city.

In notable contrast with geographers, then, sociologists have been at the centre of efforts to theorise intimacy for two decades now and competing perspectives have emerged within the discipline. The work of Giddens (1992), Bauman (2003), and Beck and Beck-Gernsheim (1995) has much in common, theorising the decline of traditional bonds of society whereby the individual becomes free to enter into and leave couple relationships as part of the ongoing reflexive project of the self. One of the most extreme predictions for the future of intimacy was made by Bauman (2003: 74), who argued that our socialisation within a market economy is 'the most awesome of dangers threatening the present form of human togetherness.' The self-interest, rational choice and competitiveness associated with contemporary consumption culture, he argued, makes people seem exchangeable or disposable altogether. Indeed, 'the man with no bonds' was Bauman's pessimistic prediction for an individualised society, whereby people become characterised by 'no bonds that are unbreakable and attached once and for all', only bonds that are 'loosely tied, so that they can be untied again, with little delay' (2003: viii). Paradoxically, however, for other commentators the freedom associated with individualisation also seems to bring a new emphasis on 'person-related stability' (Beck & Beck-Gernsheim 1995: 49) so that close relationships, commitment and attachment are understood as increasingly central to the constitution of personal identities (Giddens 1992). Perhaps, as Beck et al. argued, '[individualisation] does not mean atomization, isolation, loneliness, the ends of all kinds of society, or unconnectedness' but, rather, 'first, the disembedding and, second, the re-embedding of industrial society ways of life by new ones, in which the individuals must produce, stage and cobble together their biographies themselves' (Beck, Giddens and Lash 1994: 13).

This notion of the reflexive self is at the core of contemporary theorisations of intimacy (Oswin & Olund 2010: 60). The work of Anthony Giddens (1992) is overt in this respect. In his outline of 'the pure relationship', and drawing on therapeutic discourses in which relationships are something that need to be 'worked at' or 'worked on', Giddens (1992) proposes that 'the pure relationship' is reflexively organised on a continuous basis. He suggested that there has been a move towards relationships that are 'free-floating', unprompted by traditional obligations of social and economic life encoded in kinship and marital relations. Since commitment and reciprocity become more important than external anchors, a focus on intimacy is an expectation of such relationships, a gauge of their quality. It is not difficult to see how this links with Giddens's wider individualisation thesis in which the self forms a trajectory of development from the past to the anticipated future. In this understanding, the self is a reflexive project in which the individual is responsible for building a coherent narrative of self-identity from self-observation and self-understanding. Giddens suggests that the reflexive self must balance opportunity with risk, such that 'the world becomes full of potential ways of being and acting', a way of thinking that promotes giving up that which feels secure in order to experience personal growth.

Giddens's thesis has not been without criticism and is certainly rather abstract in formulation, yet there is something insightful, I believe, in his identification of the way in which subjectivities in the contemporary world are shifting in their emphasis towards a more reflexive approach to our intimate lives. In this analysis of the intimate subjectivities of British migrants, Giddens's understanding of 'lifestyles' is also important:

> a lifestyle can be defined as a more or less integrated set of practices which an individual embraces, not only because such practices fulfil utilitarian needs, but because they give material form to a particular narrative of self-identity... Lifestyles are routinized practices, the routines incorporated into habits of dress, eating, modes

of acting and favoured milieu for encountering others; but the routines followed are reflexively open to change in the light of the mobile nature of self-identity. Each of the small decisions a person makes every day – what to wear, what to eat, how to conduct himself at work, who to meet with later in the evening – contributes to such routines (Giddens 1992: 256).

In Chapter 4 I explore how British migrants in Dubai construct an idea of difference through lifestyle practices and narratives which help shape the cross-cultural intimacies in their everyday lives. This discussion, therefore, reminds us that any dichotomy set up between the intimate and the social is untenable.

The focus on individualisation in the work of Beck, Giddens and Bauman, as identified above, has come under attack from feminist sociology. Carol Smart (2007) points out its pessimism, preoccupation with disconnection, and the way in which it tends to ignore the ongoing significance of intimacy in contemporary life. Smart (2007) instead puts forward what she calls 'the connectedness thesis', an intervention that arises from her empirical work on accounts of personal life. While Smart is not promoting connectedness as inherently desirable, and high-lights some of the problems of connecting with others in her analysis, she suggests that a focus on connection provides us with a different sociological imagination to counter-balance the individualisation thesis:

...by an awareness of connection, relationship, reciprocal emotion, entwinement, memory, history and so on. Connectedness as a mindset encourages enquiry about all kinds of sociality and seeks to understand how association both remains possible and desirable, as well as how it may take different shapes at different times (Smart 2007: 189).

Smart (2007: 187) suggests that her focus on the sociology of 'personal life' should not be seen as an attempt to replace the notion of 'family' but, rather, is used 'as an inclusive term which does not always have family as its starting point, finishing point or as inevitable point of reference.' For me, this links to the most immediately apparent advantage of engaging with 'intimacy' as a critical theoret-ical tool: while not forcing us to abandon existing understanding gleaned from distinct sub-fields, such as notions of friendship, it allows us to examine a fuller range of intimate relationships in which we are engaged in our personal lives at the same time. Furthermore, in doing so, it encourages us to acknowledge the multiplicity of connections and attachments through which we inhabit our social lives, instead of privileging one kind of personal relationship. The term 'intimacy' thereby opens up an analytical space for us to begin to think about how different kinds of interpersonal relationships might inform each other. Of course, most of us recognise this tacitly in how we understand our own personal lives and those of others around us, but this has yet to impact upon how we organise research into intimate subjectivities. Therefore, while I organise the subsequent chapters through more conventional divisions of relationship typologies, I nonetheless remain attentive to the points in my analysis whereby our understanding of one kind of relationship is informed by another.

I have mapped the various literatures that lead to my adoption of the term intimacy in framing the empirical analysis in this book. But what of my own approach to intimacy? How might we proceed with 'intimacy' critically and, especially, spatially? In the next section, I outline a geographical approach to intimate subjectivities.

Intimacies, Subjectivities, Spatialities

There are many ways in which we might acknowledge and understand the *spatialisation* of our intimate lives. This section aims to explore this assertion and, in doing so, I seek not only to explain why a more explicitly geographical approach to studying intimacy might be productive, but also to demonstrate how a focus on 'intimate subjectivities' (e.g. Constable 2008, 2016; Mahdavi 2016) might help us to understand more fully the social geographies of everyday life.

Firstly, then, intimacy is re-shaped in dialectical relation with mobility, with practices of each informing the other, such that attention to migrants' intimate lives helps us to understand the textures of intimacy being constructed and navigated by everyone, from differently located subject positions, in everyday life. Indeed, transnational movements, journeys and connections are constructed through migrants' intimate subjectivities. I am not the first to suggest this and a body of literature on the commodification of intimacy and global work has demonstrated this extensively, through detailed empirical work on the shifting meaning of familial, conjugal and romantic relationships (e.g. Constable 2003, 2007, 2008; Pratt 2012). This is also argued by Pardis Mahdavi in her work with low-income migrant men and women across the Gulf:

> ...migration has an impact on individuals' intimate lives, and vice versa. The experience of migration, even the decision itself between going and staying, is connected to intimate ties and bonds of love. Likewise, migration has an impact on people's intimate lives. Notions of the 'family' and 'home' may change in the course of the migratory journey, and relationships between loved ones often undergo significant transformations when one or more of them are abroad for long periods of time (Mahdavi 2016: 174).

Related to this, and secondly, intimate subjectivities are co-produced in private and public life. As Oswin and Olund (2010) have argued, intimacy is co-produced within the political and economic spheres and an awareness of this helps us to move forward conceptually by unsettling dichotomies of public/private life. Oswin and Olund describe intimacy as a 'biopolitical dispositive', drawing on Foucault's definition of a dispositive as 'a thoroughly heterogeneous ensemble consisting of discourses, institutions, architectural forms, regulatory decisions, laws, administrative measures, scientific statements, philosophical, moral and philanthropic propositions' and, more specifically, its inference of the way in which these elements act in a 'system of relations' at a particular moment (ibid.: 61). While Foucault theorised sexuality (and race) as a dispositive, Oswin and Olund (ibid.) argue that intimacy also needs to be theorised as a dispositive, since, firstly, 'the nonsexual aspects of intimacy predate sexuality as the truth of the modern self, and they persist into the present', and, secondly (drawing on Berlant & Povinelli), intimacy is coming to the fore as a regulatory construct 'as the range of acceptable disciplined sexualities increases.' One aspect of this concerns queer theorists' deconstruction of heteronormativity as a set of norms governing social life. Among themselves, British migrants are rather differently positioned in relation to the production of heteronormative constructions of love and family, not least by their marital status (see Chapters 6 and 7). Nonetheless, migration reveals how constructions of heteronormativity are spatially contingent (Hubbard 2008) and then, as a result, how the particular place in which people are located becomes significant. In this case, Dubai, as a particular spatial and temporal location, was hugely important in how intimate relations emerged or not between British migrants and a range of others, including Emirati nationals and migrants positioned in varied ways through inequalities of race and class. As one illustration, Dubai's 'Decency Laws' are revealing of how state legislation can impact upon our intimate subjectivities. For British migrants, especially women, residence in Dubai requires a shift in the conduct of sexual relationships that is felt through corporeal regulation (see Chapter 6). Another example concerns Dubai's migration regime, structured by the *Kafala* sponsorship system common across GCC countries (described in Chapter 3). The refusal of citizenship to even long-term residents leads to a sense of 'temporariness' in the city, the impact of which can be traced in British migrants' narratives about friendships within Dubai and transnational connections with friends 'back home' (see Chapter 5). To fully appreciate this thread of analysis requires a conceptualisation of place as produced dialectically through both material and symbolic geographies. From this perspective, Dubai is, therefore, as much an imagined place as it is a lived place. Migrants, then, are reterritorialised in Dubai; they are emplaced, as their intimate subjectivities take shape in the global city.

Thirdly, the spatialisation of intimate subjectivities is also evident from their co-production with, and through, particular spaces, such as the domestic space of home. As Conradson and McKay (2007: 168) note, 'it is not difficult to think of work, friendship and family in these spatialized terms, for each of these forms of sociality is typically associated with a series of identifiable locations.' Geographers have already begun to explore the home as one such important site in the negotiation of intimacy, family and couple relationships (e.g. Bowlby 2011; Gorman-Murray 2006; Morrison 2012, 2012; Robinson, Hockey & Meah 2004; Rose 2003; Wilkinson 2013). Critical geographies of home very much recognise the

mutual and ongoing re-constitution of subjectivity and home, especially in relation to gender and sexuality (e.g. Blunt & Dowling 2006; Waitt & Gorman-
Murray 2011). Yet, Valentine (2008) has called for geographers to extend their
analysis of intimacy beyond the domestic and the ethnographic methodology I
used in this study identified many other important spaces in the (re)production
of migrants' intimate subjectivities in Dubai. 'Public' spaces in the way that they
are conceived in the UK are rare in Dubai (exceptions being the street and beach),
so privatised spaces of leisure figure in this book more prominently: malls, cafes,
and members-only expatriate social and sports clubs, for example. Spaces of the
night-time economy (hotel clubs and bars), for example, are identified as especially significant in the making of single British migrants' heterosexualities
(Chapter 6), an important contrast to the focus on the domestic intimacies of
married Britons embedded in marital and familial relations in Dubai (Chapter 7).
Work spaces, though also important in the production of intimacies, are not part
of the analysis (see Chapter 3 for an explanation of my focus on British migrants'
routines and practices beyond work).

It is important to resist the conceptualisation of these sites – British migrants'
homes and the privatised leisure and consumption environments they inhabit in
the wider city – as simply *proximate* spaces of intimacy. Given that these spaces can
be considered spaces of intimacy in the global city, they should be conceived of as
being trans-localities (Smith 2000), informed by, indeed constituted through,
their relationality and connectivity. This way of thinking about the relationality of
intimacy in the trans-local spaces of the global city, also resonates with broader
feminist efforts to problematise the spatial hierarchies through which intimacy has
been conceptualised. For example, while concerned with geopolitics and violence,
the work of Pain and Staeheli (2014: 345) also stresses that intimacy, 'is a set of
spatial relations, stretching from proximate to distant... a mode of interaction that
may also stretch from personal to distant/global...'. Furthermore, it supports the
theorisation of the spatialities of globalisation put forward by Amin (2002) as: 'a
topology marked by overlapping near-far relations and organisational connections
that are not reducible to scalar spaces' (Amin 2002: 386).

Once we establish that intimate subjectivities are co-constituted through place
and space, it follows that we should seek to understand intimacy as embodied.
Alongside a shift in the wider social sciences, feminist geographers began, over 20
years ago, to take the body more seriously in their work, identifying the significance of both corporeality and embodiment in our subjective sense of the world
(see for example, Butler & Parr 1999; Callard 1998; Longhurst 2001; Rose 1995;
Teather 1999). Variations in bodily comportment, practices and appearances
across space and time disrupted the idea of the 'natural' body, as well as further
critiqued the Cartesian separation of the mind and body (Longhurst 2001). The
body is important to my approach to intimacy in this book. Firstly, it is a reminder
that we negotiate embodied encounters with strangers with our identities socially
constructed and ascribed onto our bodies in ways that we cannot fully control. In

the transnational city, this is evident, for example, in the reading and interpretation of status from the nationalised, racialised, ethnicised, gendered, aged and classed migrant body. As we shall see, the segregation of Dubai is navigated through a sense of hierarchical social order, with British 'expatriate' subjectivities being mobilised in response to and through their embodied social locations. Following Doreen Massey's (1994) essay, 'A Global Sense of Place', places are seen here as sets of social relations that extend beyond the local, connecting us to people in geographically distant places. Thereby, British residents in Dubai do not simply leave their national sense of habitus behind (Edensor 2001), nor the status that their nationality confers. Rather, as transnational migrants, they continue to be informed by their identification as British and others' perceptions of them as British, seeking out, for example, compatriots with whom to join in particular performances of 'community' (Chapter 5).

The idea that Britishness is reproduced by British migrants and perceived by others as significant means that the broader social stratification of space in Dubai matters, shaping the geographies of intimacy enacted and experienced by British migrants. As has been made clear in feminist work on the embodiment of migrant subjectivities, gender, race, ethnicity, class and sexuality are co-produced in spatially contingent ways on and through migrants' bodies. But this work has primarily analysed the production and construction of marginalised subjects (Silvey 2005). Likewise, as I discuss in more detail in Chapter 3, existing research on migrant subjectivities in Dubai and across the GCC region has often focused on low-income groups (e.g. Elsheshtawy 2010; Mahdavi 2016; Mohammed & Sidaway 2016). In contrast, British migrants enact privileged 'expatriate' subjectivities, through spatialised lifestyles marked by affluence and whiteness (see also Knowles & Harper 2009; Leonard 2010). In Chapter 4, then, I explore some of the encounters British migrants have with other migrants in Dubai, including in what might be called 'intimate' spaces in terms of physical proximity (the taxi, home, or nightclub for example), but I argue that the establishment of British 'expatriate' subjectivities for the most part precludes intimacy emerging in these relations or in the city more generally. Significantly, Jamieson (1998: 1) suggests that intimacy requires equality in the relationship in which it is to be enacted: 'intimacy across genders, generations, classes and races can only take on this character if the participants can remove social barriers and transcend structural inequalities'. Furthermore, she adds (2002: 8) that close association and detailed cognitive knowledge of another are insufficient to ensure intimacy, since a degree of sympathy and empathy are key. Arguably, the politicisation of intimacy acquires further significance and new meanings in the transnational spaces of Dubai that are characterised by diversity, segregation and shifting status hierarchies in relation to citizenship.

In addition, the body becomes important to the analysis of geographies of intimacy because intimacy is an active process established and maintained not only discursively but also through relational practices (e.g. care, love, sharing, empathy) that we associate with friends, couples, and families (Smart 2007: 48). The idea of 'practices' as being key to the embodied enactment of identities is one that has seeped into our language in the social sciences, including geography and migration studies, to a point where it is now a taken-for-granted way of expressing the 'doing' of everyday life. Nevertheless, I will draw on David Morgan's (2013) efforts to locate his use of practices, since my approach draws so much on his notion of 'family practices', extending it in terms of interpersonal relationships more generally. Morgan suggests that his own use of the term is informed by Bourdieu's understanding of 'practical relationships', in contrast to the abstractions of the rules governing kinship and family. There is a link between these levels of analysis, since the rules often encode the social sense of what is the 'proper' conduct in relationships, but the practices of relationships are also 'shaped by practical considerations and immediate concerns' (Morgan 2013: 21). This is captured by Finch and Mason's (cited in Morgan 2013) idea of 'negotiation' in their analysis of how people navigate the obligations arising in family life.

The effort of negotiating interpersonal relationships certainly comes across most keenly in Chapter 7's discussion of families, where I explore the shifting gendering of home and work among married British couples. Negotiation should not be seen as carrying necessarily negative connotations here: the work of intimacy surely brings attention to its significance in everyday life. Furthermore, for Morgan in his discussion of the making of families, the attention to practices does not negate the role of discourses; rather, 'it should be clear that the two are mutually implicated in each other. Discourses are a form of practice just as practices are given meaning and shape through discourses' (ibid.: 23). Indeed, talking about sexual and romantic relationships, for instance, can be understood as a practice of friendship that can be observed among single Britons (see Chapter 6). As Morgan (2103: 26–27) argues, 'agents are frequently able to account for these practices and to recognise their taken-for-granted quality', and so 'practices of intimacy' is not a phrase I use to connote unreflexive 'doings' or interpersonal life. The use of 'practices' as a way of thinking about agency might also suggest a focus on the social actor at the expense of the structures and institutions of intimacy, yet for Morgan (2013: 24) this need not be the case, due to the clustering of actors in a 'mutually understandable set of relationships'. While 'habits' may be personal, 'the responses to them and the meanings assigned to them are more collective' (ibid.: 25) and the tension between, on the one hand, the notion of 'habits' and, on the other, the 'doings' of the social actor, both of which are traceable in the meaning of practices, is a productive one. Arguably, the recognition of the governance of intimacy (see above) further undermines any false separation of personal practices and institutions in the analysis of intimacy.

Finally, then, the notion of intimacy as embodied also further emphasises how *geographies* of intimacy are deeply implicated in our emotional lives. Valentine (2008) saw the affective register of intimate relations, as well as the concern with bodies and identities, as links between the sub-disciplinary areas of sexuality, childhood/youth, and parenting that precisely make it worthwhile to develop a geographical understanding of intimacy that can bring these debates together. Over recent decades, the discipline of geography has been characterised by relatively discrete, and at times fiercely oppositional, sets of work on 'emotional geographies' and 'geographies of affect'. Although I am not focusing so much on 'what bodies are doing', as a geographer interested in non-cognitive affectual geographies might (Pile 2010), I take the corporeality of emotion seriously in terms of the body being a site through which subjectivities are made. Bondi (2005) asserted the value of psychotherapeutic ideas in advancing scholarship on emotional geographies. Psychoanalytic accounts, she argued, 'elaborate a view of emotions as intrinsically relational as well as being intensely felt' thereby avoiding 'the twin pitfalls of equating emotions within individualized subjectivity and conceptualizing affect in ways that distance it from ordinary human experience' (Bondi 2005: 441). As such, emotions can perhaps be studied through discursive accounts, while recognising they are enacted by embodied subjects (Bondi 2005). In this book, with the focus on intimate subjectivities, it makes sense to be considering emotions in this way. This is not the approach taken by all migrationists (compare, for example, Svašek 2010: 5-8; or see Conradson & Latham 2005b), but it resonates especially with work on the emotional dimensions of transnational family life (e.g. Baldassar 2008; Baldassar & Merla 2014; Skrbiš 2008). The emotional negotiation of citizenship, racialised encounters, communities, friendship, sex, love, romance and gendered practices of parenting, work and home, are all, as I will demonstrate, part of the production of migrants' intimate subjectivities. From a geographical perspective on embodiment and emotion, then, the emplaced body is recognised as a site through which power is negotiated (Ahmed 2004) and these perspectives on emotional geographies, rather than prioritise corporeal sensation, acknowledge the significance of distanciated geographies of transnational spaces and historicity in the discursive constitution of emotion (Pain 2009). This book will demonstrate how an array of positive and negative emotions connected with embodied experiences of place (e.g. desire, disgust, love, fear) help to shape our orientations towards certain relations, and not others, partly elucidating how people make sense of the racialised segregation of social lives in Dubai. Furthermore, Pratt and Rosner (2012: 21) suggest that the emotions of the research itself might be central to the project of elucidating the global intimate, since, 'crying and laughing, it turns out, are part and parcel of feminist knowledge production and pedagogy.' As a result, I have not attempted to write myself out of this ethnographic book, however tempting that may have been.

A focus on intimacy-as-emotional also provides us with a way of understanding why the spatial dialectics through which intimacy is lived and imagined are somewhat more complex than they might at first appear. Critical approaches to intimacy dispute the equation of physical closeness with intimacy, and distance with a lack of love or care (e.g. Maclaren 2014; Thien 2005). For Deborah Thien

(2005: 192), contemporary notions of intimacy privilege 'a distinct socio-spatial character, symbolized by the open arms of the embrace,' and assume 'a distance covered, a space traversed to achieve a desired familiarity with another.' Yet, as Thien (2005: 201) goes on to argue, a critical perspective must question and overturn such assumptions: proximity to those who know us can also be experienced as constraining, and intimacy-as-distance might offer some a more flexible way of doing intimacy, 'reflecting the ambivalent and elastic spatialities of emotion.' Maclaren (2014: 56) also reflects that the meaning of 'closeness' and 'distance' in terms of intimacy 'cannot be simply literal, for one can feel profoundly close to someone on the other side of the world, and at a great distance from the person standing right here.' Furthermore, she argues, since it is intimacy with others through which our self is made, our intimate others 'are already here within us, shaping our own perceptions and experiences' (ibid.: 57):

> experiential intimacy in its most paradigmatic form involves not just a momentary inhabitation of the other's intentionality and world, but the enduring incorporation, habitual embodiment, or institution of a shared world, with the result that one's perceptions are in an on-going way not simply one's won but also informed and enabled by one's intimate others (Maclaren 2014: 60).

As a result, an intimate other can be 'both close and far, both here and there' (ibid.: 57). For migrants, it has already been established that dominant perceptions of the spatial mapping of intimacy-as-physical-closeness are brought into question by transnational practices of care between family members resident in different countries (e.g. Baldassar, Baldock & Wilding 2006; Baldassar & Merla 2014; Constable 2008). Nevertheless, the importance of corporeal proximity in sustaining transnational communities seems to be evidenced in the desire for at least occasional co-presence (Urry 2004), including the significance of 'the visit' (Mason 2004). In this book, I further explore how the geographies of home and away, here and there, mobility and settlement, inform the production of migrants' intimate subjectivities. In doing so, I also demonstrate through this ethnographic account of British migration to Dubai that intimacy is an important theoretical lens through which we can further understand the geographies of migrants' transnational life-worlds. As Maclaren (2014: 56) argues, 'most of us seek closeness with others and we do so because such closeness promises to bring with it a sense of being at home and okay in the world, a sense of belonging and mattering.' Analytical attention to migrants' intimate subjectivities reveals how they are made in relation to both mobility and settlement, unsettling the related dichotomies of home/away and global/local that obscure emplacement in their everyday lives. Attention to the geographies of intimacy of particular sites and places instead shows us that migrants are reterritorialised in embodied, emotional ways. Consequently, the lifestyle practices and subjectivities they enact shape the spatialities of their intimate practices and routines. Furthermore, critical approaches to intimacy offer an insight into how migrants are re-embedded in spaces strati-fied by race, class, gender, age and sexuality. As a result, the subsequent chapters further examine the socio-spatial stratification that has emerged in the globalising city of Dubai as the site of migrants' re-grounding and the making of their inti-mate subjectivities.

Note

1 This concern with unpicking the spatial hierarchies in how intimacy is conceptualised is also shared by Pain and Staeheli (2014) in their work on geopolitics and violence.

Chapter Three
A Globalising Gulf Region and the British in Dubai

Although Findlay (1998) identified an increase in British migration to the Middle East from the 1970s that was largely connected with the oil industry in the GCC countries, this observation was not followed up by subsequent in-depth studies either by geographers or migration researchers from other disciplines, who as a group continued to ignore the region almost entirely. Yet, the GCC region continued to grow in significance as a destination for British migrants over the subsequent decades so that, by 2005, an estimated 60,000 Britons were resident in the Emirates (Sriskandarajah & Drew 2006). In this chapter, I first examine why the GCC regions have attracted increasing numbers of migrants over subsequent decades, both British and of other nationalities. I seek to explicate the Gulf migration regimes, focusing specifically on the United Arab Emirates and, where the literature makes it possible, on Dubai itself. I focus on the meaning of citizenship and belonging in Dubai, with reference to the *Kafala* sponsorship system, the socio-cultural hierarchy that structures intimacies, and the spatial-segregation that both results from and helps reproduce this hierarchy. The spatialities of globalisation at work in Dubai take on enormous significance in this ethnographic study, as the spatial–temporal site in which the intimate lives of British migrants are unfolding. Having introduced Dubai and some of the salient factors that shape residents' everyday lives, I then go on to explain my methods and methodology for researching British migrants' everyday lives in the city. In doing so, I also turn briefly to existing literatures on British migrants to consider what we know of their existing identities, practices and experiences, especially in

Transnational Geographies of the Heart: Intimate Subjectivities in a Globalising City, First Edition. Katie Walsh.
© 2018 John Wiley & Sons Ltd. Published 2018 by John Wiley & Sons Ltd.

terms of skill, race and gender. Finally, I offer a reflexive note. As such, the chapter aims to build on Chapters 1 and 2 to further situate the subsequent empirical chapters.

Migration and Development of the GCC Region

The global significance of the Gulf states is today indisputable. No longer reliant on dazzling architectural claims, it is evidenced in their international investment strategies, trade and increasing political influence. However, the 'globalness' of the GCC region also results from the transnational connections of its millions of temporary migrant workers (Price & Benton-Short 2007). In this section, I build on the introductory discussion of the 'Trucial States' in Chapter 1, to explain how the GCC region came to have such large and diverse migrant communities. I focus this discussion especially on the United Arab Emirates, with Abu Dhabi being an oil-rich economy and Dubai developing economic diversification strategies based on infrastructure, manufacturing, leisure and tourism, free zones and real estate.

Before proceeding with this discussion, however, it is important to note that the movement of people to the GCC region is not a recent phenomenon. Tribes began to migrate to the region in the 1700s, attracted by the promise of trading prosperity. Groups that arrived at this time included the Qawasim from Persia (area now incl. Iran) and the Bani Yas from Najd area (now Saudi Arabia), settling relatively permanently in the region in increasing numbers (Heard-Bey 1997). By 1904, the population of sheikhdoms collectively known as the Trucial States included 44 tribes, with an est. 80,000 people (Heard-Bey 2001). The rapid expansion of the pearling industry also led to an influx of foreigners and encouraged permanent settlement on the Trucial coast, before the industry's collapse in the late 1920s (Heard-Bey 2004). Iranian and Indian migrations are an especially significant part of Dubai's twentieth-century migration histories. The wind towers in the *Bastakiya* area of Dubai are an architectural emblem of the long history of movement from southern Iran (Coles & Jackson 2007) and Neha Vora's (2013) research with Indian nationals also reveals a long-established community of merchants (in some cases four generations), organising trading from the Creek by dhow.

Nevertheless, the GCC region's current global significance as a dominant destination for migrant flows can be traced to the growth of the oil-rich economies. For the Emirates, as documented by Davidson (2008), the oil timeline began in 1950 when the first drilling rig was erected on the coast of Abu Dhabi to search for oil. Many dry holes were drilled in several locations across the Trucial States and offshore, but oil was eventually found in commercial quantities offshore from Abu Dhabi (1959) and in the desert (1960), export facilities built (1962) and shipments begun (1963) (Davidson 2008). Oil exploration and export led to further increases in the population, such that the first census of the Trucial States in 1968 counted 180,000 people (Heard-Bey 2004). While in the 1960s there were about 240,000 'foreign-born' among the 10 million inhabitants of the GCC countries, the number of migrants quadrupled in each of the two subsequent decades as these economies grew (Valenta & Jakobson 2016). Then, from 1980, labour migration to destinations across the GCC grew from 4 million to over 20 million in 2013 as economic transformation took place (Valenta & Jakobson 2016). This rapid increase in the number of migrants led to extreme population imbalances. Shah's (2008) report for the International Labor Organisation provides estimates for the percentage of nationals and expatriates in the population and labour force of the GCC countries in 2005, based on a suite of sources. At the time, 71.4% of the population of the UAE were expatriates and they made up 89.8% of the labour force, the highest in the region (Shah 2008). The UAE was closely followed by Kuwait and Qatar, both countries with over 80% expatriates in the workforce, and by Oman and Saudi Arabia, with over 60% (Shah 2008).

During this time, the constitution of the labour force by nationality also changed radically. Initially, labour was drawn predominantly from nearby Arab countries (non-GCC countries), including Egypt, Palestine, Jordan and Yemen. But several factors led to these countries accounting for a smaller relative share of the migration streams (Kapiszewski 2006): firstly, labour from these countries was limited in comparison with the demand of the GCC region and therefore needed to be supplemented by other source countries; secondly, during the 1990s, the Gulf war led to migrants from Palestine, Jordan and Yemen (countries that were supportive of Iraq) being expelled from Saudi Arabia and Kuwait; thirdly, the GCC authorities had raised concerns about the potential for Arab migrants to demand greater political rights and destabilise their social order; and, finally, the demand for extremely low-paid labour led to recruitment from poorer economies, especially from some regions of India, such as Kerala. While absolute numbers of migrants from nearby Arab nations have continued to increase during this time, the growth of the GCC economies has led to an increasingly diverse range of source countries. Many of these, like India, the biggest source, are in south Asia (e.g. Pakistan, Sri Lanka, Bangladesh and Nepal), but African countries (especially Somalia, Eritrea, Chad, Nigeria, Sudan) and East Asian countries (especially the Philippines, Indonesia, and Thailand) have also emerged as significant sources of labour migrants (Valenta & Jakobson 2016).

But what of the Dubai I found myself in? It is important not to obscure the diversity of either the GCC countries or the Emirates of the UAE. Differences in the precise migration histories, economic policies and transnational urbanisms across the Gulf led to city-specific geographies of settlement, belonging and intimacy. This account of migration to the region, and to the Emirates more specifically, might lead us to anticipate a British expatriate community in Dubai dominated by the oil and gas industry. Yet I did not meet a single person employed in this occupational sector until I later conducted comparative research in the region, since Abu Dhabi is the location of this industry (see Walsh 2014). British residents in Dubai instead reflect its unique economic diversification since the establishment of Federation of the Emirates in 1971. In *Dubai: The Vulnerability of Success* Christopher Davidson (2008) presented a comprehensive and convincing account of the political and economic transformation of Dubai, including this diversification strategy. Such a project is beyond my scope here, but I draw on his analysis to explain some of the most salient and necessary points to understand and situate my fieldwork.

Davidson (2008) argued that, in comparison with neighbouring Emirate Abu Dhabi – which exports 90% of UAE oil and has 10% of the world's proven hydrocarbon deposits – Dubai was not in a position to be able to depend on oil and gas revenues. While oil was finally discovered in Dubai in early 1960s, with commercial quantities found at the Fateh (Fortune) oilfield 15 miles offshore in late 1966, and Dubai Petroleum Company created to manage the oilfields, by 1991 Dubai had reached a production peak of approximately 420,000 barrels per day (Davidson 2008). Instead, therefore, Davidson (2008) documents how Dubai was forced to diversify its economy through five related strategies. Firstly, Dubai invested 25% of its GDP in commercial infrastructure, a strategy that was vital in supporting its global reach in trade relations. For example, Jebel Ali Port was established in 1979 as one of the largest in the world, and highways, bridges over Dubai Creek, and the airport have received huge investment. The second strategy combined agriculture and manufacturing to reduce import dependency, e.g. cement, plastic water bottles, workers' uniforms. The third strategy, Davidson (2008) argued, was the establishment of Free Zones, to attract foreign non-oil related investment. Evidence of this strategy is visible in the spatial concentration of particular types of company in the city. Free Zones provided infrastructure for companies and, in some cases, circumvented federal law by clustering industries in peripheral enclaves 'outside' the UAE. Davidson (2008) noted that companies locating in Free Zones were excluded from restrictions on the repatriation of capital and transfer of capital for 50 years, and free to choose employees of any nationality (except Israeli), rather than be subject to Emiratisation. While some were sceptical of these zones having any meaningful operations within the city (e.g. Davis 2005), they have continued to be successful in attracting investment. Dubai Internet City, for example, had representative offices for 100 ICT companies on opening in 2000, but by 2007 this had increased to 850 companies (Davidson 2008). The fourth economic diversification strategy identified by Davidson (2008) was an effort to focus on tourism and leisure, such that by 2000, Dubai was attracting 3.4 million tourists per year. The final major strategy that Davidson (2008) identified as contributing to the diversification of Dubai's economy was the development of real estate. He noted that until the mid-1990s there was no property market in Dubai and instead the ruling family donated land around the city to other national families to build their own houses or for residential and commercial blocks to lease to expatriates and businesses. In 1997,

Emaar Properties announced a new $200 million residential complex near to the Emirates Golf Club, to be called Emirates Hills, and advertised that they would be selling luxury villas within the development to all GCC nationals and 'other foreigners'. As Davidson (2008) notes, the launch of Emirates Hills broke UAE federal laws which stated that only UAE nationals could own land, so in the meantime Emaar offered 99-year leases to foreigners and Sheikh Muhammad promised future enabling legislation. His reassurance led to a property boom (including the Greens residential community that I toured while accompanying the Dubai Adventure Mums – a daytime social group of accompanying spouses, see Chapter 5 – one morning during my fieldwork. This diversification is a vital part of the explanation as to why demands for more varied types of skilled migration emerged in Dubai from the late 1990s and, therefore, why I encountered an occupationally diverse and increasingly class-stratified 'community' of British nationals in the city during the period of my fieldwork from 2002 to 2004, something which I detail more fully later in this chapter.

Dubai's diversification strategies sometimes overlapped in urban megaprojects that combined shopping malls, offices, leisure facilities, real estate and hotels. Many of these projects caught the attention of the global media and, arguably, further increased Dubai's attractiveness to British migrants. As Elsheshtawy explains:

> More than any other city in the region, and perhaps the world, it has made the notion of branding, i.e. developing icons which capture attention through superlatives (the tallest, biggest, etc.) or the borrowing of styles from other regions and cultures and rebranding them as its own, a key ingredient in pursuing global city status (2010: 128; see also Bagaeen 2007).

The excesses of these projects – their reliance on cheap migrant labour and environmental unsustainability – have been extensively criticised by Western commentators (see, for e.g., Davis 2005). Meaningful symbols have been chosen for the most iconic of these buildings: the Burj Al Arab is the shape of sail invoking Dubai's maritime history, while Palm Island adopts the shape of date palm fronds. Yet Elsheshtawy (2010) notes how these have nonetheless been viewed as 'artificial' landscapes and come to symbolise the domination of capital and the spatial marginalisation of low-income migrant workers. The Burj Al Arab Hotel (completed in 1999), for example, has security guards preventing access to the manmade island it is built on, 290 metres out to sea (Elsheshtawy 2010). Yet, for some of those privileged enough to have access, these mega-projects are also attractive spaces of consumption. While on fieldwork, I celebrated New Year's Eve on a newly constructed and still empty Palm Island and I must admit it felt exciting to be a resident who belonged, however temporarily, to such a moment. As the hoarding on the more recent Burj Dubai announced, construction of the mega-projects is also about 'history rising', and some have argued that it can be interpreted as an assertion of the independence of the Emirates (Stephenson 2013). Either way, British migrants and property investors have been enthusiastic in buying into these dreams, with Britons accounting for 25% of sales of villas on Palm Islands (Elsheshtawy 2010). While many residents were sceptical still at the end of my fieldwork period, this initial hesitancy had largely disappeared by the time I returned to in 2006. Scepticism appeared again after the global financial crisis, but the region was more buoyant than many commentators anticipated and demand for property certainly increased again among the British.

While migrants of all nationalities are deemed essential for the thriving economies of the GCC countries, the population imbalances between citizens and migrants have generated concerns among citizens about infrastructure demands, monitoring and security, as well as intermarriage and long-term citizenship (Kamrava & Babar 2012). As a result, the Gulf countries have established a number of policies to manage their migrant labour forces, as Kamrava and Babar (2012) identify: the requirement of temporary sponsorship for both work and residence; the segregation of both residential and commercial zones; a lack of recognition for family reunification; regular visa amnesties encouraging migrants who have over-stayed visas illegally to return to their homeland; nationalisation of the workforce through quotas; a rotational labour system and the diversification of the nationalities of workers to reduce the likelihood of organisation. I turn now to consider the first of these policies, the *Kafala* sponsorship system, since it underpins the entire structure and implementation of this strict migration regime.

Citizenship, *Kafala*, and the Temporary Migrant Worker

In Dubai, as well as across the wider Gulf region, the distinction between citizen and expatriate is key to understanding the legal and social status of residents and their socio-spatial separation in everyday life. Fargues (2011: 278) describes the Gulf states as having 'dual societies', without the new, mixed population we might otherwise anticipate to emerge over time from the coexistence of nationals and migrants:

> *De jure* separation is embodied in all the legal provisions that differentiate between nationals' and foreign nationals' rights and duties, the obligation of every foreign national to have a national sponsor (the *Kafala* system), the prohibition of intermarriage (with only a very few exceptions), and foreign nationals' lack of access to a number of family rights (such as family reunion and access to public education for their children) and labor rights. De facto separation is reflected in the lack of or in the severely limited interaction between nationals and non-nationals. The two groups have some interaction in the workplace, but even this is limited because labor markets are highly segmented.

While obscuring the diversity among both citizens and migrants, this dichotomy does have resonance with the UAE context. Since there are no formalised processes of naturalisation, the citizen/expatriate distinction persists irrespective of the length of residence of a migrant in the Gulf. Indeed, in his study of second-generation migrants in Dubai, Syed Ali (2011) suggests that the Emirates has been very strict in granting citizenship since the 1990s. The requirements are unspecified, but Ali (2011) reports that they are thought to be: over 30 years' residence, being Muslim, of Arab descent, and an Arabic speaker, as well as having a clean police record, good academic qualifications, a certain level of bank savings, and a degree of personal influence (*wasta*) to lobby a sponsor. Instead, across the Gulf and irrespective of length of residence, work visas are linked to a *Kafeel* or sponsor, a UAE citizen, typically an employer or, in the case of entrepreneurs and business owners, an investor. For Buckley (2012: 255), 'The UAE's *Kafala* rules most principally reflect rulers' efforts to balance the need for large numbers of foreign labourers against the desire to discourage and restrict foreigners from settling permanently in the country,' but is also 'linked to a host of ongoing tensions and concerns among local rulers about national security, cultural preservation and rising unemployment among Emirati nationals (Kapiszewski 2006)'.

The negative impacts of the *Kafala* sponsorship system on migrants in the Gulf are now widely reported and understood, especially with regard to low-income migrant workers from south Asia (e.g. Buckley 2012; Human Rights Watch 2006). Alongside other critical scholars, Michelle Buckley (2012) argues that the *Kafala* rules offer employers the power to limit or withhold wages, and to extract illegal fees from migrant workers to cover such things as driving licences. Recruitment is subcontracted in efforts to circumvent the UAE federal labour laws and the Keralite construction workers Buckley (2012) interviewed typically took between one and three years to repay the loans their families had used to pay recruitment agencies for visas and plane tickets, frequently stating that they were misled about the salaries they would receive. However, Buckley's (2012) research, taking place after the 2008 financial crisis, reveals another huge advantage of the *Kafala* system to GCC employers, namely the flexibility of labour to employer demand, for it allowed the immediate and large-scale deportation of migrant workers, without compensation, when finance was withdrawn from many of Dubai's construction projects leading to their postponement or cancellation. Thereby, Buckley (2012) argues, much of the risk of these kinds of speculative real-estate projects associated with Dubai's urbanisation is carried by the migrant worker rather than the company employing them.

The *Kafala* sponsorship system also has an impact on migrants' intimate lives. A non-working spouse and dependent children can be sponsored by the working migrant only if their salaried income is above the minimum threshold required for a family visa, making it impossible for low-income migrants to have their families accompany them in their migration to the Gulf (Gardner 2011). Male 'bachelors' are consigned to living in labour camps and marginalised in poorer city districts, such as Deira, but their physical exclusion does not stop them from being considered 'a menace to local society' by citizens, who view them as 'a threat to their family, to their personal security, and to the integrity of their culture' (Gardner 2011: 19 drawing on Shah). While all migrants are considered a cultural threat, 'bachelors are feared for their potentially out-of-control sexuality that is believed to endanger women and family life and, as such, the social order' (Mohammad & Sidaway 2016: 1405). Smith (2010) documents the 'bachelor fear' in Dubai, where a harassment telephone line was established (after my own fieldwork had ended). Female domestic workers are less visible and not demonised to the same extent, but their reproductive lives must also be enacted in the homeland, in common with transnational domestic workers in other regions (e.g. Pratt 2012). Mahdavi (2016) documents how domestic workers who become pregnant outside marriage in the UAE risk their own imprisonment and the prolonged statelessness of their child. The British families I discuss in this book are, therefore, a marker of the privilege of these migrants: the higher income of British migrants enables them to bring with them their spouse and dependent children. The additional costs of private schooling and housing, as well as the limitation on sponsoring elderly relatives, leads to nuclear families being the norm even among middle-class south Asians (Gardner 2011).

The impact of the *Kafala* system on middle-class skilled migrants, such as Britons, clearly does not lead to employment precariousness or infringement of reproductive choices to the extent experienced by other migrants. Nonetheless, both of these factors emerge in their narratives and impact upon their intimate subjectivities, as I will demonstrate through later empirical chapters. Furthermore, as a result of the denial of citizenship, arguably migrants across all skill levels in the Gulf economies share a particular mode of being perhaps best characterised by the phrase 'permanent temporariness' (Bailey et al., cited in Mohammad and Sidaway 2016: 1410). For Mohammad and Sidaway (2016: 1410), the *Kafala* system generates 'the rhythms' of low-income migrant workers in Qatar, who must live the 'forced temporality' of 'punctuated lives', either working intensively to save up for return and marriage, or sustaining a 'bifurcated work and family life,' coming and going over decades. Likewise, Elsheshtawy (2008b: 972) suggests of migrant residents in Dubai: 'the general sense for any member of its long-term population is that he or she is a visitor – somebody who eventually leaves regardless of how long they have stayed'. As Ali (2011: 557) reminds us, *all* visa categories in the UAE are temporary and the lack of state incorporation – 'defining them as temporary migrant contract workers, rather than as immigrants' – compels transnational orientations, even among a middle-class second generation. Ali's (2011: 562) research reveals how this group honour a 'social contract' in return for the tax-free 'good life', giving up any claims for permanence and living 'fully prepared to leave'. The possibilities of extending and securing settlement in the UAE through property ownership were just beginning to be discussed among British migrants during my fieldwork as a way of circumventing the *Kafala* rules. It was not until 2006, however, that Dubai passed a law confirming the right of non-citizens of the UAE to own real estate in certain developments built since, such as the Palms, Jumeirah Islands, Emirates Hills, the Meadows, and Arabian Ranches (Bagaeen 2007). While this legislation removed some of the uncertainties and risks of property investment, ownership does not automatically confer residence rights, even now. Preoccupation with the notion of 'temporariness', whether in relation to their own migration trajectory or a generalised transience among other residents, came across strongly from British interviewees, even those who had lived in Dubai for over twenty years. They often spoke of 'the firm' rather than the government being in control of their movement and uncertain futures, and the idea that Dubai is host to a 'transient' population was a dominant theme. This transience often emerged as a perception that was deemed to negatively influence everyday practices of intimacy among Britons, in spite of the many advantages of 'expatriate' life. The impact on intimacy is evident in British migrants' accounts of friendship (Chapter 5) and romance (Chapter 6) in particular.

The UAE's migration regime and labour market, organised around the *Kafala* system described here, leads not only to a rigid sense of social hierarchy between nationals and migrants, but also to a complex cultural politics reflecting the stratification of urban society along intersections of class, race and nationality. I explore this further in the next two sections by, first, describing the social stratification and, second, by discussing wider research on Gulf migrant subjectivities.

Social Stratification and Migrant Subjectivities in Dubai and the Gulf

Dubai has been described as 'a place of and for mobility' (Junemo 2004: 183) and Benson-Short, Price and Friedman (2005) identified it as the city with the highest proportion of foreign-born residents worldwide (82%). UAE Ministry of Labour statistics from 2005 demonstrate that flows from certain regions dominated the migrant labour market at the time of my research (Elsheshtawy 2010: 212): resident nationals from India (51%), Pakistan (16%), Bangladesh (9%) and Sri Lanka (2%) made it a largely south Asian population but, as remains the case, Arab non-GCC nationals (11%) and migrants from the Philippines (3%) were also relatively significant. Europeans and US nationals accounted for less than 2% of migrants according to these figures, with British residents too few to list separately. In the first broad review of migrant labour in the Gulf, Kapiszewski (2001) described the way in which various migrant groups were differentiated by status in Dubai in a way that resonates with the city I encountered in 2002 to 2004: at the very top of the social hierarchy were Sheikhs and their families, followed by other Emirati nationals of Arab or Persian origin settled over several generations. Wealthy Arab immigrants from Iraq, Palestine and Egypt, some of whom have been given Emirati national citizenship, were identified as the 'migrants' with most social status. It is worth noting that from the perspective of British migrants, however, these distinctions among those they identify as 'Arabs' are not frequently known or easily identifiable. The middle strata of the social hierarchy is occupied by other wealthy and professional migrant groups, including upper and middle-class Indians, Lebanese, Iranians, and Europeans, including the British (Kapiszewski 2001). As I will demonstrate in Chapter 4, however, British discourses frequently contest this through a sense of cultural superiority. During my fieldwork, and certainly since, it became apparent that affluent Russian, Chinese and African migrants were increasing in number and adding to the diversity of this social strata. The lowest social position in the hierarchy as described by Kapiszewski (2001) is inhabited by a large migrant population from South and South-East Asia, predominantly India, Pakistan, Sri Lanka, Bangladesh, Thailand and the Philippines, working in what are deemed 'unskilled' or 'low-skilled' jobs, especially in the construction, hospitality and retail industries, and in domestic work.

When considering the racialisation of Dubai's urban spaces, the complexity of the social hierarchy makes it impossible to think through a black/white binary. The society is 'kaleidoscopic' in Malecki and Ewers's (2007) terms. Indeed, even to list these national groups in succession obscures the complicated positional hierarchy that exists within these broad groupings. While employment opportunities (and the associated wages, lifestyle and status) in Dubai are linked to nationality, variations also exist within national groups, based on class, occupation, religion, education, caste etc. For instance, Neha Vora's (2008, 2013) research introduces us to Indian working, middle-class, and elite migrants, employed in a wide range of occupations, whose level of education, English language proficiency and socio-economic background grants them access to higher-wage jobs, as it does in other global cities, complicating the picture of the Asian-migrant-as-construction-worker reproduced in the global media. While Vora (2008, 2013) reports racial discrimination operating against middle-class Indian migrants and describes their experiences of status negotiation as non-citizens, their experiences are of course not comparable with those of many Indian nationals who work in the construction industry alongside other south Asians (see Human Rights Watch 2006) or are isolated behind closed doors in domestic work (Sabban 2004). British migrants in Dubai are also diverse and I explore this in the next section, but first I further explore Gulf migrant subjectivities from existing research, focusing the analysis on their socio-spatial lives, belonging, and intimacies. As Malecki and Ewers (2007: 477) suggest, the segregation and polarisation observed previously in global cities 'takes on new forms in the Middle East'. In this section, then, I examine how fears about mixing across class and religious/cultural lines inform the spatial segregation of migrants' lives, while paying attention to the diversity of Gulf migrant subjectivities as they are themselves diversified by ethnicity, race, class, gender and generation.

Probably the most well-known example of the segregation of space across the Gulf is the housing of low-income construction workers in labour camps. These camps are located either in the industrial zones of cities or far beyond the cities, necessitating daily bus transportation, sometimes for several hours at each end of the working day. Sonapur, 15 kilometres from Dubai, is the location of some of the most infamous of these labour camps, condemned by the international media for their high-occupancy, unsanitary conditions and lack of facilities for workers (e.g. Human Rights Watch 2006). These labour migrants are geographically isolated by their housing, poverty and the lack of transport infrastructure between the camps and the city. Walking is discouraged in Dubai by the huge seven-lane highways that both channel and control mobilities: 'Roads become walls, boundaries and lines to be navigated, alienating pedestrians' (Kendall 2012: 46). Male construction workers do walk, but they must navigate slip-roads, illegal entry and exit points, hoardings, fences, construction sites, empty lots, and the sun to break these 'spatial taboos' (Kendall 2012: 51).

Beyond the extremes of the labour camp, urban space is further organised along lines of affluence and race. Much of what we understand about low-income migrants in the UAE has come from Yasser Elsheshtawy's extensive research, informed by urban theory, on the neighbourhoods of 'old Dubai' – Deira, Satwa and Karama – places often 'conceived of as poor and dangerous places filled with "Indians"' (2008b: 976). In spite of their marginalisation, Elsheshtawy's (2008a, 2008b, 2010) ethnographic research in these south Asian neighbourhoods suggests that low-income migrants make efforts to claim comfort, inclusivity and belonging in Dubai. In an effort to look beyond the urban spectacle of Dubai's mega-projects, Elsheshtawy highlights the active and vibrant street life of these neighbourhoods where low-income residents gather, especially on a Friday (the weekend). The areas are in fact ethnically diverse, especially Karama, with Indian nationals joined by residents of other nationalities in the lower-middle strata of Dubai's social hierarchy, particularly Filipino, Iranian and Lebanese migrants. Elsheshtawy (2008b) argues that such residents sustain their transnational connections through the everyday practices they engage with in these neighbourhoods: emailing relatives in the internet cafes, eating familiar foods in ethnic restaurants, transferring remittances using the money exchange operators, and shopping in the supermarkets. However, friendships and socialising with their compatriots is often the biggest pull, with migrants gathering in public space 'at the edge of large spaces, empty lots, green lawns, parking lots, and next to traffic lights...incidental spaces that were not planned for such use' (Elsheshtawy 2008b: 984–985).

In contrast, Vora (2013) describes how a minority of elite Indian residents have chosen to move to newer developments in Dubai, while others in lower-paid professions with families have been pushed outwards to suburbs, where housing is cheaper, or even to neighbouring emirate Sharjah. Nevertheless, the historical and persistent spatial concentration of Indian nationals in these areas led to some of her interlocutors to describe Dubai as 'India's Westernmost state'. In response, Vora (2013: 66) argues that:

> Indians experience Dubai not as a hybrid elsewhere, but as a space infused with Indianness and not fully distinct from the subcontinent, and this is particularly true of the Indian-dominated neighbourhoods in old Dubai. India is generated and lived on a daily basis in old Dubai, as is evident in the texture of daily life within its neighbourhoods and in how these spaces are – and are made into – distinct sites of urban citizenship.

Like Elsheshtawy (2010), Vora (2013) is informed by a critical project to uncover the more ordinary and everyday claims to urban space which negate, to some extent, the legal exclusions Indian residents face from their formal designation as 'temporary' migrant workers. As such, Vora (2013) argues, the citizen/non-citizen divide is made more complicated, with Indians narrating themselves as outsiders with regard to Emirati nationalism, but with emerging subjectivities that make claim to their place of residence through affective belongings. Those who belong to the Indian merchant class, for example, not only participate in the *Kafala* system as managers and employers, too, but draw upon a nostalgic 'frontier' narrative that locates them firmly in the illicit trading history of the Gulf when their business expertise was regularly consulted by Sheikh Rashid and the ruling family. Many of these families have also lived in Dubai over two, three or even four generations. Furthermore, Vora (2013) suggests that young Indian nationals born in Dubai are increasingly staying on to attend higher education institutions alongside Emirati nationals, rather than returning to India or studying in Europe or the US and Canada. This is resulting in new subjectivities: 'South Asian diasporic youth narratives were beginning to resemble liberal class for rights and equality, and were pushing again their sense of being "second-class" citizens in the only place they knew as home' (Vora 2013: 34).

While British residents would seem to share the position of Indian middle-class nationals in Dubai's social hierarchy according to Kapiszewski's (2001) depiction, their experiences are in fact very different. Indeed, their racialisation of and by others, and the making of white 'expatriate' subjectivities in the city, is discussed in detail in Chapter 4, since it is so important in understanding their intimate lives. In the next section, however, I introduce the British 'community' in Dubai.

Researching British Migration in Dubai, 2002–2004

Using ethnography to research expatriate belonging meant that I spent approximately seventeen months living in Dubai. Ethnography enabled me to be flexible and spontaneous in my research approach, although I incorporated several easily identifiable strategies. When I first chose Dubai as the location for my research I had never visited the city myself and knew no-one else who had either. In fact, none of my friends or family had ever heard of it, something which seems inconceivable in 2016. Nevertheless, I was convinced by an initial web-based research that it was a suitable choice for doctoral research, not least because there was a vibrant expatriate community and it would be a safe country in which to undertake research as a single woman. I did an extensive literature review on the UAE so that I would be aware of the society within which I was preparing to live. I also made a two-week preliminary visit in an attempt to avoid previously unforeseen problems and focused on practising 'hanging out' in expatriate spaces and thinking through the practical and logistical elements of living in Dubai such as accommodation, grocery shopping, transport, internet availability, and appropriate clothing. This helped me to be confident about my ability to live there and gave me some contacts to reach out to in the first days of my long-term stay.

At the beginning of my fieldwork period, I began by conducting interviews with 'gatekeepers' from British expatriate society, including: representatives from British institutional groups such as the British Community Assistance Fund (BCAF), the British Business Group (BBG), and the British Embassy; community groups and social societies which British expatriates belong to such as the Dubai Adventure Mums, Mother-to-Mother Breast-feeding support group, Dubai and Sharjah Women's Guild, Dubai International Arts Centre, St George's Society, and the Welsh Society; key individuals in the 'British community' such as head teachers; the managers of several British pubs; editorial staff from expatriate media including *Time Out Dubai, What's On, Living in the Gulf, Explorer Guidebook, Stars Magazine, Identity Magazine*; and finally, experts from several relocation companies and estate agents who help affluent migrants with their move and orientation in Dubai.

Simultaneously, I immersed myself in participant-observation of British migrants' everyday lives in Dubai, from as many different access points as I could, although with varying success. This included routine attendance at a number of 'expatriate' spaces and groups, including: the Women's Guild and Dubai Adventure Mums; a course of riding lessons at the Dubai Equestrian Centre; fitness classes; a course of evening classes in Arabic, and some in photography and pottery, as well as an excursion with the Dubai International Arts Centre; sessions with Dubai Archery Group.

To supplement this participant-observation, I used in-depth interviewing strategies, by which I mean repeated interviews of an hour or more duration and 'shadowing' of everyday activities. Having initiated such a range of contacts through the strategies of participating directly within the British migrant community in as many places as I could gain access to, I managed to later recruit a wide range of interviewees to work with in a more sustained and in-depth manner. The purposive sampling I employed in this research was important in capturing the diversity evident in their 'community' in Dubai, a diversity that resonates with other recent studies of British migration (Knowles & Harper 2009; Leonard 2010; Scott 2006). Importantly, in Dubai, British migrants are employed in a diverse range of occupations. My interviewees therefore included male and female lead-migrants working in a range of sectors associated with highly skilled migration: engineering, project management, finance and law, and entrepreneurial businesses. However, British graduates also worked in managerial positions in the retail and hospitality sector, and in positions in the health, education, media, recruitment, marketing, tourism and public relations sectors. British migrants also work in non-graduate jobs as, for example, airline crew, hairdressers, bar staff, and administrators, while women were occasionally employed in low-paid positions as teaching assistants and retail assistants. As such, categorising the British migrant community in Dubai through existing typologies is not easy. The community certainly includes a 'transnational elite', in terms of individuals that frequently relocate within transnational corporations – Beaverstock (2005: 249) described this group as 'highly-mobile, highly-paid and highly-skilled' – but considerable variation exists in terms of length of residence (between 1 and 25 years for my participants), travel biography (Dubai was the first international relocation for some of my participants while others had moved frequently), and benefits package (housing and education allowances, for instance). This diversity was fully represented in my ethnography, alongside age, marital status and household composition (co-habiting single migrants in apartment or villa shares, as well as married couples with and without children living at home). As a result, this research challenges an implicit conflation of the terms 'expatriate', 'highly skilled', and 'British migrant'.

Interviews were individually tailored to the participants, but they each began with more generalised questions to ascertain background information, for instance on biography and reasons for coming to Dubai. One of the subsequent interviews was directed at domestic material culture, so was based in a participant's home and included a tour of the house and conversations around moving home and particular 'things' or 'biographical objects' (Hoskins 1998). I also conducted ethnographic shadowing with these interviewees, something which allowed a wider range of concerns to emerge unprompted. This meant gaining access, by invitation, to a further range of informal occasions, such as nights out in bars and clubs, bingo at a golf club, quiz evening at a sailing club, Christmas carol evening at The Dubai Country Club, group excursions, wadi-bashing, a national's wedding, the Islamic Centre, barbecues, trips to the beach and many more occasions that I would not otherwise have been able to participate in. Having made brief notes or mental notes where more appropriate, I later wrote about all these places, the events, what happened and what people did and said, in my ethnographic diary. Participant-observation was perhaps the most important part of the ethnography in terms of allowing for the findings on intimacy to emerge as significant. I did not design an ethnographic study to explore intimate lives so, however useful it turned out to be, I do not put this forward as the model for researching intimacy and, indeed, other qualitative and creative methods are equally productive (see Gabb 2010; Morrison 2012, 2013; Smith 2016). In addition, however, I also conducted many singular and informal 'interviews' with people who were not willing or could not commit to longer periods of involvement in the study, thus maximising participation through the flexibility of an ethnographic approach. With longer-term contacts I initiated these informal interviews to follow up something happening or being said during my participant-observations. In an effort to make these 'interviews' relaxed, spontaneous and convenient, I conducted them anywhere, including by the pool, over coffee, and while a participant was driving. Despite this disparity in location, these quotes are labelled here as 'Interview' and were recorded with consent and transcribed. In addition, I include field notes from conversations and observations made in the course of participant-observation. I followed a technique of establishing an open and interactive dialogue, making it clear that nothing was irrelevant to my interest. For ethical reasons, I was always honest about my position as a researcher and gained verbal consent to accompany interlocutors for fieldwork purposes and to record interviews. In addition, I have made every effort to make the people whose words or actions are included in this thesis anonymous: names are changed, some details are changed where necessary (e.g. occupational details), and no photographs are included. The British migrants who became my close friends *do not* feature in this book, but inevitably inform my interpretation and understanding of the material I gathered. The last part of this chapter provides further details on this distinction, by way of a reflexive account of the fieldwork.

The research focused on domestic and social spaces, with relatively little engagement with migrants' working lives. In part, this focus was because access to working lives was difficult to achieve. Exceptions included shadowing Stephanie for a day as she gave guided tours on an open-topped bus and also Jane on a typical day working as a hairdresser. Although several male participants invited me to their workplace when I requested an initial interview, these visits were brief and involved meeting with the interviewee in a meeting room rather than watching them as they went about their usual work. This has the unfortunate consequence that I could not include ethnographic material with regard to how British expatriates might relate to other expatriate communities and Emirati nationals in the workplace, something that would be a useful addition. In fact, I experienced some difficulties in gaining access to men, especially married men, due to time constraints (as discussed, men are more likely to be employed outside the home), attitude (as they were often the sole wage earner, married men tended to designate domestic space as 'the wife's department') and a relatively low level of 'natural' rapport arising from own gendered positionality. My family based work, for instance, was generally initiated through contacts at the Dubai Adventure Mums and Dubai Women's Guild, so tended to rely on the invitation of women. Where possible I extended this to talk to the husband and/or children of a family, but this was often difficult due to their commitments to work and school, and because people were generally reluctant to give up their evenings (see Valentine (1999) on the practical problems as well as power dynamics of household research). However, my ethnographic research was also deliberately less concerned with working practices, since previous research on expatriates and skilled British migration has largely focussed on their working identities (Beaverstock 1996, 2002; Beaverstock & Bordwell 2000; Findlay et al. 1996; see also Leonard 2010 and Leggett 2013 for a focus on expatriates at work informed by postcolonial theory). A focus on work would necessitate obscuring other practices that are also central to expatriate lives, such as those connected to domesticity, sociality and leisure, here incorporated by my focus on intimacy. Furthermore, in a society where many British women do not work, expanding the focus away from work-places and working identities also has the positive effect of increasing the inclusion of women's experiences and voices.

Research on couples participating in highly skilled migration has consis-tently identified the persistence of more traditionally gendered divisions of labour within households, irrespective of nationality and geography of migra-tion (Coles & Fechter 2008; Hardill 1998; Kofman & Raghuram 2005; Yeoh & Willis 2005). Hardill (2002) demonstrates that many highly educated British women with considerable work experience and established expertise end up moving as a 'trailing spouse'. She suggests that married women continue to face significant obstacles to becoming a 'lead' migrant, including gendered power relations in the household and gendered institutional norms in transnational corporations. Furthermore, once committed to accompanying their partner, many women find it impossible to begin or sustain careers in situations of circulatory or successive relocations (Hardill 1998). In Dubai, as we shall see, there are also single British women migrants (see also Fechter 2007; Willis & Yeoh 2007), nevertheless, barriers to work exist for women whose move is initiated by their husbands' employment, include visa restrictions, language, invalidity of qualifications, and impermanence. Since a move to Dubai usually increases the disposable income of the family, even while they shift to a single income, and the wages of migrant domestic workers are kept low, such house-holds can afford to employ part-time or full-time help with cleaning, cooking, gardening, and childcare around the home. Therefore, many women become removed from the paid labour force, but this does not necessarily infer a domes-tication of their lives. Chapter 4 reveals the increased opportunities for leisure enjoyed by many women in these circumstances during the daytime while their children are at school. Chapter 7, meanwhile, discusses in more detail how shifting gendered subjectivities of both men and women can bring challenges to intimate couple relationships.

A Reflexive Note

Reflexivity – defined as 'self-critical sympathetic introspection and the self-conscious analytical scrutiny of the self as researcher' (England 1994: 82) – is deemed to be a critical and productive part of the ethnographic research process (e.g. Bell, Caplan & Karim 1993; Silvey 2003). It might also be understood as especially important in researching intimacy (e.g. Pratt & Rosner 2012). As such, I use this final section of the chapter to provide a reflexive note before presenting my empirical analysis in Chapters 4 to 7. In doing so, I draw largely on the more immediate reflexive account I produced for my doctoral research during the year following my return from fieldwork in Dubai, but supplement this by further awareness that has arisen in the drafting of this book over a decade later. While for some readers this section may raise questions about the objectivity or reliability of the research, I take this risk as a vital part of the story of this book since, as Gillian Rose (2004) has argued, knowledge is always situated and partial. Indeed, in Rose's (2004: 247) words, we 'see the world from specific locations, embodied and particular, and never innocent'. As a result, 'feminist geographers should keep those worries, and work with them' (ibid.: 259) since,

> We cannot know everything, nor can we survey power as if we can fully understand, control or redistribute it. What we may be able to do is something rather more modest but, perhaps, rather more radical: to inscribe into our research practices some absences and fallibilities while recognising that the significance of this does not rest entirely in our own hands (Rose 2004: 260).

Ethnographic research is perhaps more demanding of reflexive accounts than other methods, with authors producing highly personal, honest, and emotional descriptions of the challenges they experienced in their research (e.g. De Lyser & Starrs 2001; Hobbs & May 1993). Nevertheless, I do not provide a general account of the fieldwork here but instead, given the focus of this book, highlight a discussion of field intimacies (see also Smith 2016).

Many rather different intimacies are central to ethnographic research, evoked in distinctive contributions framed with the terminology of rapport, friendship, sexuality and love. A common concern, however, centres on the impact that intimacies have on the production of an effective ethnography. 'Successful' participant-observation, for instance, 'relies on the researcher's ability, at one and the same time, to be a member of the group being studied and to retain a certain detachment' (Denscombe 1995: 144; see also Pearson 1993: xviii). Distance, movement, and boundaries remain central preoccupations in ethnographic field cultures, despite the turn towards reflexive geographies which deconstruct such authorial claims. Such concepts persist in descriptions of the skills of the 'good' ethnographer as someone who: develops rapport while practising detachment (Bernard 1994: 136–137); navigates 'insider' and 'outsider' positions; and someone who more often than not moves 'away' to research and comes 'home' to write 'the scholarly equivalent of war stories' (Katz 1994: 68).

The majority of authors reflecting on the power that shapes their intimate relationships in the field do so in a context of researching a subordinate or marginalised 'other' (e.g. Butz & Besio 2004). The power relations were rather more equal in my research but, nevertheless, the establishment and maintenance of relationships, at the centre of ethnographic fieldwork, brings responsibilities and obligations, whomever the subject and whatever their status:

> we can never not work with 'others' who are separate and different from ourselves; difference is an essential aspect of all social interactions that requires that we are always everywhere in between or negotiating the worlds of me and not-me [...] we are never 'outsiders' or 'insiders' in any absolute sense (Nast 1994: 57).

While I have mobilised theories of whiteness in the analysis of the privilege I observed in Dubai (see Chapter 4), for ethical reasons I made every effort to protect the anonymity of interviewees in my decisions about how to present the findings, as well as ensuring I did not treat people as 'research objects' (Stanley & Wise 1993). In an attempt to give something back to the 'expatriate community', I decided to help with Rainbows (a Girl Guiding group for 5–7 year olds) each Wednesday afternoon. Despite learning much from my participation and initially presenting a paper on my experience, I decided it was unethical to include within my thesis or publications since I did not initially approach the organisation with the plan to do research, though they knew that I was a researcher. Of course this volunteering was not a direct swap of time and energy with the many more British residents who helped me. As feminist geographers have noted, it is the researcher who, through the advancement of their career, often gains most reward from the research experience (e.g. Katz 1994).

Developing friendships within Dubai heightened the significance of ethical decisions, decisions that must be made whatever the chosen method but that seem especially salient in ethnographic research. Many of these arise from the ways in which everyday life in the field undermines the distinctions often drawn in the construction of knowledges:

> Where are the boundaries between 'the research' and everyday life; between the 'fieldwork' and doing fieldwork; between 'the field' and not; between 'the scholar' and subject? (Katz 1994: 67).

The people I shared a villa with often joked that doing a PhD seemed to be a 'complete doss' and, catching up in the evening, they might sarcastically ask: 'Been working hard today Katie? How many coffee mornings did you go to?' or 'Interviewing by the pool? You must be exhausted!' Comments like these reassured me that my position as a researcher had not been forgotten, but also revealed that my housemates did not understand 'research' to cover my relation-ships with them at all times. As a result, I have respected these boundaries and included discussions with my housemates only when I was able to mark them clearly as research (by recording them as interviews or by seeking permission to overtly record field notes). To protect their privacy, to maintain the trust of my friends, and also to give me time-off mentally, not every moment could be treated as fieldwork.

As many ethnographers have argued, there are some relationships that develop during fieldwork that feel like genuine relationships, beyond the doing of research and from the researchers' perspective. Here the possibility of the ethnographer 'belonging' is raised, such that, 'participant observation becomes almost indistinguishable from living the culture; ethical dilemmas dissolve; and possibilities for tracing the rhymes and rhythms of the society in an ethnographic text seem boundless' (Bell 1993). I, too, felt these comfortable moments of belonging. However, rather than helping ethical dilemmas to dissolve, it seems to me they actually complicate ethical responsibility in a different way. It is not only the ethnographer who can perceive the genuineness of honest connectivity so it is precisely at such times of forgetfulness that others are likely to tell us things or let us observe things that it might not be appropriate to include in a responsible ethnography. Indeed, as a result, my closest friends do not appear at all in this book. However, it is inevitable that a significant amount of my under-standing came from repeated and long-term interaction with my friends, as I became part of their lives. The events we shared and my 'knowledge' about my friends cannot simply be eradicated from memory, and therefore it informs the text indirectly.

Ethnographic research has been described as a 'way of life' rather than a job, due to its extensive, personal, emotional and sometimes traumatic nature (Armstrong 1993). I lived in Dubai for only a short period in comparison with other places I have settled, but it has shaped my own geographies of intimacy, belonging and home immeasurably and in ways I do not pretend to fully under-stand. As an ethnographer doing fieldwork, I hung out with British residents in Dubai, the people who became interviewees, interlocutors, and friends. I ate, drank, danced, played, exercised, emailed, watched TV, shopped, sun-bathed and travelled, with British migrants and, perhaps increasingly, as a British migrant. Because I lived in Dubai, I established intimate friendships in Dubai: I laughed, chatted, shared my past, and planned my future alongside British migrants, as an 'expatriate' to a certain extent. For a while at least, I lived the British 'expatriate' positionality, albeit subjectively from my own embodied perspective. On return, I experienced intense feelings of being out of place socially and in relation to various academic communities. I now recognise this as a common symptom of long periods of international fieldwork, having witnessed it in others, and these feelings were an understandable response to my return. I experienced a grief of sorts: certainly, deep regret for a life I was actively choosing not to live in Dubai, sadness at having left behind significant friends, and a sense of home-sickness. Working out how I came to feel at home in Dubai was a key part of sorting through the ethnographic material for the original analysis in my thesis and led to my resistance to stereotype Dubai and its residents.

The flexibility that sometimes becomes necessary in the field might also be understood as one of ethnography's strengths: when open-ended, projects might grow and define themselves (Silvey 2003). In this way, the project became one of researching migrant homemaking differently, through attention to intimate subjectivities. As I researched migration, I disrupted my own sense of home. I cried real tears at Heathrow airport and, eighteen months later, at my leaving party, surrounded by new friends in Satwa. I suppose, really, I simply learnt for myself some of the complex relations between migration, intimacy and home. The next chapter is the first of the empirical chapters that aims to elucidate not my own experience of these relations, but those of the heterogeneous British migrant 'community' I encountered in Dubai.

Chapter Four
British 'Expatriate' Subjectivities in Dubai

Its cultural landscape has seemingly blurred the distinctions of East/West, orient/occident, colonial/post-colonial, etc. Yet a closer examination reveals a movement towards segregation and fragmentation, a post-urban centre resulting in a new mix of urban spaces (Elsheshtawy 2010: 4).

As Yasser Elsheshtawy (2010: 3–4) points out, 'a new form of urbanity is emerging' as Dubai responds to its positioning as a gateway between the urban centres of the East (Shanghai, Singapore, Hong Kong, Mumbai, etc.) and the 'crumbling, perhaps even decaying, urban centres of the West'. Elsheshtawy's (2008a, 2008b, 2010) explication of this distinctive urbanity focuses on the everyday lives of low-income male migrant workers from south Asia, based on his close observations of their encounters and practices in the multi-ethnic high-density neighbourhoods of Deira, Satwa and Karama in Dubai. Here he finds the traces of their transnational lives and celebrates the conviviality through which low-income migrants resist both their lack of citizenship rights (irrespective of length of residence) and their marginalisation from Dubai's emerging landscapes of consumption, based on mega-projects that mix residential, hotel, retail and office space. In contrast to these low-income bachelor migrants whom Elsheshtawy (2010) discusses, British migrants have a relative freedom to navigate the city, arising from their rather different positioning in terms of nationality, race and class. In this chapter, I explore the cross-cultural encounters of British migrants in Dubai and the role these encounters (or lack of them) have in the making of British 'expatriate' subjectivities.

Transnational Geographies of the Heart: Intimate Subjectivities in a Globalising City, First Edition. Katie Walsh.
© 2018 John Wiley & Sons Ltd. Published 2018 by John Wiley & Sons Ltd.

Before discussing the spatialisation of British lifestyles in Dubai's urban fabric (in the last part of this chapter), I first explore how their segregation from both Emirati nationals and low-income Asian migrants comes to be understood as 'natural' through the making of 'expatriate' subjectivities. I show how the complex socio-spatial stratification of Dubai's society means that these 'expatriate' subjectivities are both locally inflected and resonate with global articulations of being a privileged migrant. The chapter starts by considering how British migrants understand Emirati culture not through meaningful intimate relationships with Emirati nationals, but through touristic and bureaucratic encounters framed by notions of 'culture shock'. It continues by exploring the ways in which British migrants encounter low-income Asian migrants in both mediated commercial relations and public space. Finally, it considers the everyday spatialised routines through which Britons inhabit Dubai's 'landscapes of privilege' (Conway & Leonard 2014). Although, as Chapter 1 suggested, I am not alone in trying to understand and analyse the making of British 'expatriate' subjectivities, in this chapter I engage with, and extend, this existing work through my concern with the spatialisation of globalisation and intimacy. The tacit and overt notions of boundaries and respectable relationships that British migrants use to navigate intimacies in Dubai's public and private spaces in the city are key, I shall argue, to understanding how their friendships, couple and family relationships are enacted.

British Imaginaries of Dubai and Emirati 'Culture'

The everyday lives of the British migrants I interviewed in Dubai were, for the main part, characterised by minimal interaction with Emirati Gulf Nationals and a high level of interaction with other privileged migrants, including fellow nationals. Evidence from a recent survey of Western expatriates in the UAE suggests that this remains the case, with minimal interactions between Emiratis and Westerners, even in the workplace (Harrison & Michailova 2012). Emiratis represent less than 20% of the overall population in Dubai and are very poorly represented in the private sector labour force which employs most skilled migrant workers (Kapiszewski 2001), so many Britons simply do not meet Emiratis in their workplaces and, where they do, the Emiratis will usually have much higher social status (as detailed in Chapter 3). Therefore, although in fieldwork I heard of exceptions, close collegial relations between Britons and Emiratis were extremely rare. Furthermore, couple relationships and marriage were also becoming increasingly unlikely. Though I could not gain access to their perspectives directly, I had been made aware by several interviewees that there existed a small number of British–Emirati couples who had met in the 1970s and 1980s when some elite Emirati men attended military training and higher education in the UK. Yet, since the early 2000s, opportunities for education have increasingly been sought in the US or at home in the UAE and perceived threats to the cultural heritage through which Emirati national identity is constructed, as well as exclusionary citizenship regulations, means that such relationships are actively discouraged. The lack of intercultural relationships between Britons and Emiratis is not, then, explained by language barriers. Although few Britons speak or make the effort to learn Arabic (an exception being those who are schooled in Dubai, for whom it forms part of the curriculum), the younger generations of Emiratis have often been educated in English. Meanwhile, though Emirati citizens and British migrants share 'internationalised' spaces of consumption (e.g. hotels, cinemas, malls, restaurants, coffee shops), intimacy – either in terms of friendships or couple relationships – is rarely initiated through these kinds of privatised consumption spaces in Dubai, as in other global cities (Wilson 2012).

The lack of a more general social interaction with Emirati nationals and their cultural life tends to disappoint Britons who may embrace migration not only in terms of opportunities for career progression or an increase in salary, but also to encounter cultural difference. Michaela Benson and Karen O'Reilly (2009) suggest that lifestyle migration is a relocation in which people are searching for a 'better way of life'. Although they limit their definition to being within the developed world, they note the significance of a 'historical continuation of earlier mobilities', a narrative of escape, and position lifestyle migration as part of the reflexive project of the Self (Benson & O'Reilly 2009: 608–609), all of which resonates with how we might understand the way in which at least some Britons approach their residence in Dubai, even though they would typically be framed as economic rather than lifestyle migrants. Most notably, perhaps, while cautioning against a reduction of lifestyle migration to tourism, Benson and O'Reilly reflect upon how these mobilities are related:

> Tourism, of course, is based on all those distinctions Urry (1990) recognised between leisure and work, home and away, everyday and holiday. It is about escaping the drudgery of the routine in order to 'gaze' on the exotic and other; the perfect foundation for an anti-modern migration in search of community, security, leisure, and tranquillity.

> The pursuit of a better way of life that characterises lifestyle migration is the tourist's pursuit of authentic experience (MacCannell 1976) made epic, an embedded feature of daily life within the destination [cf. Benson 2007; O'Reilly 2007b] (Benson & O'Reilly 2009: 614).

In this section of the chapter, then, I explore how these themes of travel emerge as part of the varied performances of 'expatriate' identities of the British in Dubai.

There is evidence of a curiosity among the British migrants towards their 'hosts', as illustrated in an email extract that was circulated from a friend who worked in the tourism industry:

> Hello All! Now you are always asking me about 'local' weddings and having the opportunity to see something that not everyone gets the chance to see…well…this is your chance!!! Believe me you will be disappointed if you miss it! (field notes).

While economic opportunities are often paramount, British migration to Dubai is partially motivated by, and frequently conceptualised through, discursive scripts in which a particular geographical imaginary is constructed. Arguably, this imaginary draws upon a nineteenth-century romantic mindset in which travel has been constructed as an end in itself: a form of learning and pleasure (Duncan & Gregory 1998). As such it relies upon operation of certain colonial tropes, including Orientalist discourses informed by desire (Said 2003) and a form of time–space substitution whereby the Other becomes located in the past (Fabian 1983). In this persistence of an industrialised romanticism, 'travel is still popularly understood as the immersion in picturesque, distinct, colourful cultures' (Duncan & Gregory 1998: 8), so that many British migrants desire to encounter cultural practices and performances that are noticeably, sensually, distinct from their own. The opportunity that travel brings for cultural learning was certainly important in some stories of residence in Dubai, as Claire (late thirties, mother) explained:

> I thought we'd get more absorbed into the UAE and Dubai...that's the point of being abroad... Part of living abroad, well for me, is that you learn something about the culture of where you are living (interview).

Claire was in Dubai with her two children and husband, a computer analyst, and it was her first experience of international relocation, but this was also a sentiment shared by those who had previously lived abroad. Joanne (twenty-nine, public relations in Media City), for example, had had an 'expatriate' childhood, being born in Japan and living there until she was seven, and then moving to Dubai, Portugal and Holland with her parents. Back in Dubai with her boyfriend, she explained:

> The other thing I don't like about Dubai is the lack of culture here. I've always loved travelling; I've always liked going and meeting the people and throwing yourself into the culture, but there isn't any here. They hide it behind a great big Western front and that annoys me. I love walking next to the Creek: you can see the old cobbled streets, you can feel the Arabic, and you can feel the history there (interview).

Likewise, Greg (mid-thirties, recruitment advisor) claimed a highly mobile lifestyle:

> We lived in Turkey and Bahrain before Dubai. I adored Turkey, Istanbul. There's so much culture. It's true of Tokyo to a certain extent although it's still a new city, recently built, but there's culture underneath. It drives me nuts that everything outside your door here is a shopping centre. I loved Bahrain: it was so much less developed (field notes).

In the Western imagination, there is a long history of the Middle East being characterised not only through immutability and exoticism, but also its disillusioning reality (Said 2003).

A theme that emerges from these comments is a contradictory one: on the one hand, British migrants consider that Dubai's modern city was built too quickly to be accepted as genuine and yet, simultaneously, that this same modernity undermines their perception of the authenticity of 'Arab culture'. As Alistair Bonnett (2000: 79–80) argues, in the 'West' the desire for primitivism is produced within modernity as a 'flight from whiteness', so that 'the more non-western, the further away and the less modern, the better'. Thereby, the traveller is always belated (Behdad 1994; see also Fabian 1983; see Walsh 2011 for a fuller discussion in this context). The comments of Matt (late forties, engineer) and Tom (early fifties, banker) were typical and illustrate this well:

> This culture doesn't fascinate me…what's left of the old? They're ruining it all! Yes, they have to keep up with the world, but I think they're taking it to extremes, they're taking it too far. The way they're building it up, it'll just be a glorified holiday place eventually (interview).

> There's one thing I don't like about Dubai and that's all the high-rises. It's a shame they've tried to make it look like a Western city, like an American city. They could have incorporated a far more Arabic sense of the region (interview).

Such narratives, of course, reveal more about British migrant's expectations of the Middle East and understanding of 'culture' than about the Gulf nationals themselves. As Stephenson (2013) notes in his analysis of Dubai's tourism strategy, the ultra-modernisation of the city is a challenge to efforts to develop an understanding of indigenous forms of heritage and tradition, for migrant residents, tourists, and even among the younger generations of Emiratis themselves.

British residents sometimes, where possible, attempt to engage with the symbolic practices they understand as being of Emirati culture. However, this is often through more general notions of Middle Eastern culture, for example, with Emirati cuisine yet to be commodified in restaurants, eating Lebanese food was often labelled an 'Arabic take-away' (see also Stephenson 2013). In their search for an 'authentic' Arabia, I witnessed British migrants embracing tourist practices, independently and through tourist operators. For instance, longer settled migrants, especially those with families, might take excursions into the desert by 4x4 for a day-trip or weekend of camping, or visit surrounding countries, especially Oman, that are viewed as 'unspoilt' by modernity in comparison. Others, especially single and younger migrants, went on excursions organised by tour operators, often south Asian, accompanied by their visiting relatives and friends or alongside other tourists. These tours were based on the kinds of 'ludic forms of enjoyment' that Stephenson (2013: 730) identifies as most important to the sector, such as camel-riding, dune-bashing and sand-boarding. He suggests that 'the tourism script not only disengages with culture, but also potentially derides the prospects of appreciating indigenous life' (Stephenson 2013: 730). I noted, however, that these trips also appear to fulfil an imaginative desire to view a 'timeless' and 'exotic' Arabian culture, for instance the villages of the Mussandam Peninsula or the souk in Muscat are highlights. During an organised desert tour, migrants or their visitors would be exposed to the 'light Orientalism' of tourist encounters across the Middle East, including belly dancing, shisha smoking, opportunities to dress up as Bedouins, henna tattoos, and a camp fire (see Haldrup & Larsen 2009). Knowledge of the Emirates thereby becomes mediated through south Asian tour operators and the mobilisation of stereotypes through which spaces of the Orient are produced by the traveller (Haldrup & Larsen 2009). It is clear that such encounters have nothing to do with Emirati nationals themselves and work to further entrench citizen/migrant distinctions.

Occasionally, residence brings opportunities for cultural engagement and understanding that are unavailable to tourists, yet remain touristic in their framing by migrants. Field notes from my first attendance of a Henna night (part of the celebration of marriage) accompanying Stephanie (late twenties, tourism/ marketing) illustrate how positive intentions can frame these touristic encounters yet, at the same time, how they can be undermined by the persistence of Orientalist tropes around dress, gender/sexuality, and modernity:

> Stephanie is so happy and excited in anticipation, you could mistake her for the bride. When we arrive we are awed by the palace draped in thousands of white fairytale lights. We make announcements and take photos of ourselves continually: 'This is unbelievable!' 'This is amazing!' We feel embarrassed in our baggy outfits with long-sleeves, chosen (with ignorance) for their modesty. Although we are aware it is a women-only event, we are surprised that only a few older women are wearing Abayas, others in long low-cut, sleeveless, evening dresses. Afterwards, Stephanie comments: 'it's nice to see the women having a good time. It's changed my view on the way they live: they have quite a good life (field notes).

In contrast, Josh (early thirties, journalist) had lived in Dubai for six years and held an apparently more sophisticated understanding of 'culture'. Telling me a story about young Emirati nationals talking secretly by mobile phone from different tables in a cafe, Josh strongly disputed the prevalent notion of the Emirates having no 'culture':

> Many expatriates have no interest in their environment. I studied the politics of the Middle East at degree level, so I know what I'm doing here from an intellectual point of view. I need that intellectual stimulus. As a resident here, though, you have to be more proactive, ask, look around, read it for yourself, but it makes me more interested in everything that goes on here. What's the point of living here if all you want is sun, sea, sand and sex? You might as well live in Marbella. The whole region is fascinating. It's because people have a false understanding of modernity, it irritates me (interview).

Josh's fascination translated into the sense of attachment necessary for him to continue living in Dubai and a commitment to doing so. In the years following this period, he relinquished the company position that brought him to relocate and instead developed a successful property business.

Kathy (early sixties, entrepreneur and artist), a long-term resident of Dubai who arrived in the mid-70s with her husband and young children, also demonstrated a more sophisticated understanding of culture in her interviews. For example, her nuanced understanding of cultural practices in the Middle East came through in her comment, 'Emirati culture is not something you can put in a nutshell,' in explaining the differences between typical Lebanese dishes and 'traditional' Gulf food, in describing changes in Bedouin geographies over time, and telling histories of the Bastakiya heritage area. Kathy also socialised a lot with Emirati nationals, through friendships developed with other artists, in the business community, and through her sailing club. Her comments revealed affection for the differences she perceived:

> My philosophy in life is that a) you never stop learning and b) your personality is in a constant state of evolving as well. I didn't set out to read about the Emirates, it has been very much a process of daily contact. It's changing so much: The Emirates that I came to is not the same as it is now. The essence is the same. The nationals' approach to life is that same, absolutely wonderful, 'Why not?' philosophy, as opposed to 'Why?', always looking for opportunities, and I think that probably comes from the fact that life here was survival mode…they're risk takers. It's in the nature of how the Emirates has always survived: what could be riskier than being a pearl diver or being a merchant who relied totally on the wind to take a dhow over to Mozambique; they were people who took those risks armed with knowledge about the environment they lived in (interview).

Kathy described herself as living a 'twinned life' between the UK and Dubai, but spoke of her 'abstract' sense of belonging to Dubai and missing Dubai when visiting the UK. She had also been involved in preserving the windtower heritage in the Bastakiya area, contributing to a sense of emplacement for Kathy.

There is clearly evidence of a diversity among the British in terms of how they imagine and understand Dubai and the Emirates through notions of temporality and culture. Consequently, their performances of 'expatriate' subjectivities are varied. The next section further explores their understanding of difference and culture, this time thinking about how guidebooks and embodied experiences of migrant status help to shape their production of 'British' subjectivities in opposition to an essentialised 'Arab' citizen.

'Culture Shock' and Emotional Resources of Whiteness

In the absence of interpersonal relationships with Emirati nationals, British residents' views of the Emirates and its citizens were formed not only through their tourist practices described above but also their encounters with the national legislation and bureaucracy. It is in these encounters that they gain a heightened awareness of their lack of citizenship and the implications of their migrant status. Vora (2013) has suggested that Indian business elites in Dubai have become quasi-citizens, who are satisfied by their rights to generate surplus and wealth from lower-paid migrant workers. Britons, too, are for the most part seduced by the comfort afforded by their expatriate packages and their high status relative to other migrant groups. However, there are occasions when they become affronted by the way in which they are marked as migrants and their lack of citizenship removes certain rights. In this section, I demonstrate how these embodied experiences are key sites in which migrant subjectivities are produced per se, as well as being significant in the demarcation of nationality based subjectivities.

As outlined in Chapter 3, UAE legislation firmly distinguishes between the rights of citizens and those of migrants. For example, all migrants in the UAE must have a national partner to sponsor their businesses through 51% ownership and they are not allowed to purchase property beyond select developments. The residency visas of migrants are dependent on their full employment (or that of their spouse), although the introduction of property ownership legislation since 2001 has removed this requirement, making it possible to consider retirement in Dubai. Residency visas are also dependent on passing a health examination, including a screening for HIV. Through such legislation, Britons gain a sense of their migrant status not only on arriving, but periodically through their years of residence when, for example, they apply for a new visa or similar documentation. Britons frequently cite a causal link between experiences of legislative control or bureaucratic inconvenience and their own opinions of Emirati nationals, using discourses of their own 'culture shock' in explanation. But what is 'culture shock' and why should we be critical of these explanations?

The guidebook *Culture Shock! United Arab Emirates* is an example from the worldwide series of resident guides and defines the term in the following way:

> Culture Shock is the stress that occurs when people try to cope in adjusting to the newness of the environment, culture, and people. The stress can feel like an illness and can cause physical problems such as allergies, backaches and headaches or interrupt the normal digestive processes [...] It can bring on emotional stress expressed through unexpected and inexplicable tears or vehemently expressed anger over minor offences (Crocetti 2000: 43).

In their empathy towards migrants experiencing 'culture shock', such texts implicitly mark their reader as migrants with European heritage:

> Each person arrives in the Emirates with their own special set of expectations. As these are not met, the newcomer becomes intolerant and expresses annoyance in little outbursts of rage. Such phrases as, 'This would never happen in my own country,' 'These people are so backward, why can't they get it right?' and 'Arabs are crazy!' dominate the conversation of the culture shock sufferer (Crocetti 2000: 45).

Binary ontologies of difference are often used to explain emotional experiences of residence away from home as 'natural'. For example, in the guide *Don't They Know It's Friday?* the author asserts: 'The profound differences between the way of life in the Gulf and that in the West can lead to severe culture shock' (Williams 1998: 19–20). Given that bureaucracy is difficult in any unfamiliar context and migrants in Britain report much worse problems, it is evident that Britons' experiences of it may relate to their perception of the superiority of Britishness in relation to 'the Arab world' and the automatic rights they feel this should confer. All these guidebooks were (and remain) prominently displayed in English-language bookshops in Dubai.

In Dubai the embodied-emotional experience labelled as 'culture shock' and 'stress' in these books is mainly discussed in direct relation to the negotiation of formal government institutions and bureaucracy, since, as noted above, British migrants' encounters with Emirati nationals are otherwise minimal or structured by workplace regulations. To prepare the English-speaking migrant for being a non-citizen, *The Dubai Zappy Explorer*, for instance, welcomes readers with the following statement:

> The Zappy Explorer endeavours to make sense of Dubai's perplexing and mind numbing bureaucratic everything! In this tedious war of papers and permits, we provide you with the ammunition to pull through unscathed [...] Even Dubai 'old-timers' sometimes find themselves struggling to keep up with the latest rules and regulations, not to mention fads and fashions [...] So here it is – the Zappy Explorer: your best chance for navigating through Dubai's administrative mysteries (Explorer Publishing 2002: iii).

With its appeal to the need to forcefully regain control over the migration environment by making it legible, this introductory statement reflects the notion among migrants that Dubai is unfamiliar and inhospitable. Furthermore, the unfamiliarity is understood as being a result of illogical processes – 'fads and fashions' – rather than government changes in policy that might be expected in a rapidly globalising city with associated social, economic and physical transformation. Similarly, a spokesperson for the Explorer publishing company explained why these kinds of publication are considered necessary:

> Anything that you actually need to do in Dubai involves reams and reams of paperwork, and huge amounts of queuing. The way the culture is organised means that there are people who will take your piece of paper and look at it for twenty minutes, stamp it, and then pass it on to somebody else, and you'll queue for another twenty minutes, they ask you to get it translated into Arabic three times, and then someone will ask you for twelve passport photos which haven't been mentioned before, and then they'll ask you for a hundred [UAE] Dirhams, when it was twenty yesterday! The whole culture here is not as efficient as you'd expect in the UK (interview).

In spite of international problems with the bureaucracy of immigration, British migrants' discussions of the bureaucratic procedures in Dubai focus on the irritations as being attributable to the 'culture' of the Emirates.

Unfortunately, these emotion-laden discourses do not stay contained within texts or the conversations of their authors, but instead circulate as wider discursive resources in the making of whiteness and, more specifically, the construction of binary positions of expatriate/citizen and Britain/the Emirates. As an illustration, these field notes highlight the way in which Sally (late fifties, entrepreneur) discussed her visa application through which she applied for residency through the *Kafala* sponsorship system:

> Normally it is pretty straightforward, but suddenly they've changed the rules! It's the same thing with all this bureaucracy: your mind has to be blanked against it. I go robot-like through the procedure, blood tests what have you, and if it all gets done without any hassle then I consider it a huge bonus. The real wind-up is that you can't keep up with the changes. You just never know. You think you have it all sorted out and then some smug bugger behind the counter says: 'no, no, law changed'. My friend's sister speaks Arabic and she says she heard a 'dish-dash' (a derogatory term for male Emirati national) behind the counter saying, 'what can we do today to shake things up?' It's as if these people decide to be objectionable! You know, they have this brand new, state-of-the-art airport that cost trillions of Dirhams, but the stamp they put in your passport says 'Valid Thirty Days' when they actually mean sixty! It's a third-world country behind the gloss (field notes).

Unlike Sally, Joanne (late twenties, public relations) had only lived in Dubai for two years, but she also considered herself experienced as a migrant (having previously lived in many countries and describing her life as 'very expat') and shared a similar sense of shock on realising the impact of migrant status in the Emirates. When I interviewed her, Joanne was about repatriate to the UK with her boyfriend. He was going to be deported, having served a three-month sentence for breaking (allegedly) the zero-tolerance drink-driving legislation of the Emirate:

> And now, unfortunately, I'm so anti-Arab and I'm stereotyping and labelling, which, I never thought I was that sort of person but they just annoy the shit out of me. This is their country, but the thing that gets me is they promote it as this '*honey pot*', and they're building all these marinas, and they want it to be the international sort of jet setters' paradise. Well, they can't have it both ways. I don't mind paying tax [in the UK]. I like the fact that (pauses, laughs) it's a *normal* country. You don't have to pay taxes here, but you pay for it in other ways. We pay taxes in the UK, but we do have a very civilised regime there: if anything happens with the police, I know they're trained to look at things properly and there are laws that will protect you if you're innocent, or not. I like the fact that we have *proper* police and *proper* laws in the UK (interview).

In Joanne's account, the urbanisation of 'new' Dubai again becomes framed as deceptive, with Emirati nationals ('they') accused of building a modern city and tax-free haven that isn't matched by the legislative norms of everyday life. Tim Edensor (2001) argues that a key process in the 'becomingness' of national identity abroad is precisely this drawing of boundaries between self and 'other', enabled through reductive and essential notions of who 'we' are compared with 'them'. In her narrative Joanne is, therefore, engaging in the *production* of Britishness. Edensor (2001) suggests that boundary consciousness may become more pronounced away from home, sometimes suddenly or unexpectedly, since we lose competencies we previously took for granted,. Indeed, Joanne legitimated her slippage from Gulf nationals to 'Arabs' by reference to her anger: suddenly she was reminded of being a deportable migrant worker. Sally, too, was frustrated by the bureaucracy surrounding her migrant status. Both Joanne and Sally spoke with disgust of the injury incurred by, and victimisation of, the British migrant community in a way that resonates with defensive uses of hate in contexts of anxiety (see Walsh 2012 for a fuller discussion of this). Ahmed (2004: 43) argues that in the UK, hate is mobilised in relation to immigrants and 'such narratives work by generating a subject that is endangered by imagined others whose proximity threatens not only to take something away from the subject (jobs, security, wealth), but to take the place of the subject'. While the direction of migration is here reversed, and with it the majority/minority position, Gulf nationals are nonetheless portrayed as threatening towards the privileged status of white Britishness through their legislative and bureaucratic control over migrant status, residence and mobility. British migrants experience this as a disruptive episode, precisely because it interrupts the everyday articulation of a privileged migrant status, one that is related to an imperial articulation of whiteness (Fechter 2005; Leonard 2008; Yeoh & Willis 2005). This, I would argue, is the 'undeclared history', as Ahmed (2004: 47) puts it, of public emotions of anger and frustration through which 'the Arab' and 'Arabic culture' becomes saturated with affect. As a result, the negative emotions through which British 'expatriate' subjectivities are partly made might be understood as investments in the social norms of whiteness (Ahmed 2004: 56). Fortunately, in the postcolonial city of Dubai, these emotions do not lead to violence, discrimination or marginalisation in terms of relations with Gulf nationals.

It is important to note, too, that there are some British migrants who resist the articulation of British 'expatriate' subjectivities by contesting these 'culture shock' discourses. Angela (early thirties, personal assistant), for example, described how she came to Dubai with her husband, a pilot, with a plan to stay seven years (in order to qualify for a tax-free payout in lieu of pension contributions) but, once they had lived there for three years, she told me, they 'just knew' they would not be going back to the UK for a lot longer. Angela had also observed the social segregation of British migrants from Emirati nationals, but spoke of her attempts to use the Explorer guidebook to learn more about the culture. She 'confessed' to feeling 'a blind fury' in the first few months when it felt as though they were being 'duped', but differentiated herself from the 'shocking reputation' and 'embarrassing behaviour' of Brits abroad and was instead empathetic to the 'closed community' of the Emiratis ('they're not museum exhibits'). Rather than blame 'the culture of the Emirates', Angela instead drew on her psychology degree to explore British migrants' experiences of residence in Dubai:

> It's me that's changed. As humans, we're very adaptable and resilient. Now, [the bureaucracy is] water off a duck's back. But for some it's just too much, they can't cope and leave. We knew a couple who moved here at a similar time and they left after eighteen months: he went into a deep depression. He was like a spoilt child and couldn't cope with it. He was a typical Type A personality: short temper, imposing, domineering (I'm focusing on the negative traits) high octane, the sort of person who when he's in a traffic jam has his foot on the accelerator, and has his keys out three streets before he gets home (interview).

Angela, like some other British migrants, resisted the Othering that seems to act as a discursive resource in the making of white 'expatriate' subjectivities. Instead, she described Dubai as 'home', evoking the image of waking up to the sun coming through her bedroom window in a 'welcoming' way. In spite of having been previously employed in social work in the UK, however, she spoke about feeling intimidated on returning for visits: her 'horror' at seeing 'beggars, homeless people, [I shouldn't say it] junkies, everyone tattooed, smoking, over-weight' (interview). Clearly, then, while resisting the racialisation of British 'expatriate' subjectivities, Angela is happy to identify, nonetheless, with the affluence of British migrants in Dubai. Arguably, the 'landscapes of privilege' (Conway & Leonard 2014) she now inhabits not only map the spatialisation of class segregation in Dubai, but also help to reproduce it. The next section explores these landscapes further.

The Spatialisation of an 'Expat' Lifestyle

So far, this chapter has detailed how British migrants rarely interacted with Emirati nationals. Instead, they found themselves unsettled both by the unexpected modernity of Dubai's landscape and consumption practices, and by the bureaucratic and legislative reminders of their lack of citizenship and hence barriers to long-term residence. As a result, Britons drew on the resources of whiteness and affluence, the opportunities afforded to them by the residential zoning and privatised leisure spaces of this postcolonial city, to assert their privilege within the city and counter their non-citizen status. In doing so, they were able to differentiate and spatially demarcate their own 'expatriate' subjectivities from those of other 'migrants', especially low-income migrants from south Asia. In the making of their everyday lives through particular spatial resources, it is possible to observe the making of race (see Conway & Leonard 2014: 82 who argue this in relation to South Africa) and its entanglement with class. Perhaps the most visible reminder of this in Dubai's urban landscape are the exclusive gated communities and distinct residential neighbourhoods in which wealthier migrants often live, though their choices as to which malls, supermarkets, cafes, bars, and leisure facilities to frequent are also highly significant. In this section of the chapter, then, I map out the spatialisation of lifestyle and whiteness in the everyday lives of British migrants in Dubai.

As noted in Chapter 3, at the time I conducted my fieldwork in 2002–2004, the suburban residential districts where non-citizens can now buy property were only just emerging. As such, the residential neighbourhood of Jumeirah was still the neighbourhood most associated with Britons and, to a lesser degree, Europeans more generally, since it became popular in the late 1960s. Indeed, this neighbourhood was key to the making of British 'expatriate' subjectivities as both a material and discursive resource. A residential district on the coast, characterised by low-rise, detached villas with gardens, it was described by property listings magazine *Better Homes* (even in 2010, when residential options had hugely increased) as 'one of the more expensive and exclusive areas in Dubai'. Although many are original properties, older and smaller than newer villas, the small roads and coastal location remain covetable.

This area was also home to one of the first Spinney's supermarkets, the grocery chain that was the most popular with British migrants because it carries branded products from Tesco and Waitrose throughout the year and specialist seasonal foodstuffs for Easter and Christmas. Budget-conscious British families also used the Park N Shop and Union Coop, but they did so more discretely. Prior to the later explosion of massive international shopping malls that were being built towards the end of my doctoral fieldwork in early 2004 (e.g. Mall of the Emirates, Ibn Battuta, and Dubai Mall next to the Burj Khalifa), Jumeirah was also one of the routine destinations for consumption in the semi-open-air Jumeirah Centre, Jumeirah Plaza, and Palm Strip shopping centres. At the time, these were the location of reputable nail bars, hairdressers and beauticians, the English book shop 'Magrudy's', private medical clinics, and European-styled patisseries and cafes. The name of the district became part of a popular term used to describe non-working migrant women, the 'accompanying spouses' of the expatriate community: 'Jumeirah Jane' (see also Walsh 2007). At best, this is a gentle caricature of women who spend their days at leisure. Yet the term is also used maliciously to hint at the dark side of expatriate life, inferring that such women have a dependence on prescription drugs to treat boredom and depression, indulge in personal grooming, and seek the emotional and sexual attention of their personal trainers or tennis coaches. Such terms reveal the ambivalence of British migrants towards the excesses of 'expatriate' subjectivities they are offered by their privilege and the gendered dimensions of their roles and associated stereotypes.

Mercato mall in Jumeirah, as well as Dubai City Centre mall, the BurJuman Centre, and Wafi mall, were spaces that British migrants used alongside affluent migrants of other nationalities and Emirati nationals. Inside these malls were international brands of coffee shops (e.g. Starbucks), restaurants (e.g. Paul's) and high-street shops. British chains were well represented among the clothes shops especially; BHS, Marks and Spencer's and Top Shop were located in different malls and acted as a draw to British women of different ages. During the daytime, malls were largely empty and, in a city for the most part un-walkable due to the road systems and heat, they provided an alternative space in which to exercise. Importantly, however, the malls also segregate Dubai's wealthier urban residents from south Asian low-income migrant workers. For Ara Wilson (2012), the mall provides a space in which public life becomes privatised and, therefore, more highly regulated and restricted. In her discussion of intimacy as a category of transnational analysis, Wilson (2012: 36) identifies shopping malls and gated communities as architectural forms that draw on the symbolism of older forms of town-based public intimacy arising in pedestrian-friendly streetscapes and squares, yet which 'exemplify commercially produced modes of selective, priva-tized public intimacy'. Encouraged by the extremes in climate, car dependency, and global architectural norms of city-building, these kinds of privatised spaces dominate Dubai and are the site of everyday intimacies beyond the home. Through this organisation of space, capitalist market and property rights are used to construct a controlled mode of intimacy (Wilson 2012). Malls in Dubai are no exception and they enable heightened possibilities of the regulation of intimacies. As such, privatised leisure spaces serve to both maintain the distinction between citizen and migrant, affluent and poor, as well as produce differentiated migrant subjectivities. Low-income migrants are excluded from shopping malls by 'a variety of cues and measures which communicate that they are not welcome' (Elsheshtawy 2010: 973). Furthermore, Friday, likely to be migrant workers' only day of leisure, is often designated 'Family Day' at shopping malls (as well as in public parks or beaches), thereby prohibiting their entry (Gardner 2011). As a result, Kanna (2014) argues that the privatisation of leisure spaces in postcolonial Dubai allows for 'cultural exclusivism'. British migrants either do not acknowl-edge this or perceive it as 'natural'. In Chapter 5, I extend this discussion by exploring the members-only clubs and social institutions through which 'expa-triate' subjectivities are further made.

The choices of British migrants as to where to live, shop, hang out, and how to travel help to distinguish them from low-income Asian migrants and, in doing so, contribute to the making of 'expatriate' subjectivities through privilege. Britons make themselves at home in this city by inhabiting Dubai's privatised shopping malls, hotel restaurants and bars, and sports and health clubs during their leisure time. These are the 'landscapes of privilege' that Conway and Leonard (2014) describe as being important resources in the construction and performance of white British identities in postcolonial South Africa:

while difference is produced through multiple social mechanisms of South African life, it is the materiality of space that makes its daily operations starkly visible. Exploring who owns what, who goes where, how this or that person travels and why, helps to expose the texture of racialised privilege and its performances (Conway & Leonard 2014: 80).

When British migrants try to resist the reproduction of 'expatriate' subjectivities by doing what they might do at home and leaving 'expatriate' spaces, it does not always prove successful. Encounters with low-income migrants in public spaces are sometimes unsettling. The following extract from my field notes shows how the experience in public spaces of observing the difference of others, as well as being marked by their own whiteness, helped to further encourage segregated geographies of the city. In the extract, I describe an attempt by two relatively newly arrived British skilled migrants I was shadowing to use a public strip of beach at the edge of Jumeirah, a neighbourhood that is otherwise considered expatriate friendly:

The beach is very busy, so we walk for a while to find an empty spot and then lay down our towels. A group of about thirteen south Asian men – Indian nationals perhaps – in their early twenties are walking along the beach. Several walk in pairs and are holding hands. Although they are still fifty metres or so away, several are watching us already. They are wearing ties and black trousers, and their shirts are tucked in. Nick snorts with laughter: we are in swimwear and he considers them overdressed for the beach. The low-income migrant workers walk past, their gaze oriented towards us. Rachel sighs. 'They're all perverts!' she states, closing her eyes to ignore them. 'They're so gay!' Nick mutters, staring defiantly back. The group of men look back towards us, continuing to watch us as they walk away. Eventually Nick cracks and shouts: 'No, you're the strange one!'

No longer able to relax, we leave the beach. We head for the Lime Tree café, a popular 'expatriate' cafe where we will 'see some normal people', Nick suggests. Since it's just across the road we walk, but the taxis beep their horns constantly to attract our attention. Nick starts getting angry and waving his arms to gesture them away. 'Slowly they're making me hate them,' he tells me. 'I'm beginning to understand why the people who've been here for ten years treat them like shit.' 'Like being racist?' I ask. 'Yeah,' Nick continues: 'Soon, I'll be saying I don't care if you're from fucking Kerala, just fuck off! Like I used to think it was sad when an Indian treated me with too much respect, now I just think they're all...' Drinking cappuccinos on the terrace, it's notable that everyone relaxes, feeling safe from the surveillance of the beach and street beyond (field notes).

Evidently the racialisation of British 'expatriate' subjectivities through whiteness does not always make British migrants' negotiation of public spaces in Dubai a comfortable or easy experience from their perspective, something that resonates with the instabilities of white expatriate identities that has been observed more widely in postcolonial spaces where the 'Other' looks back (Fechter 2005). As a result, some neighbourhoods, such as Deira and Karama, examples of those inhabited with large numbers of migrants from other countries, are rarely visited by British migrants. Nevertheless, they do encounter low-income migrants in everyday life as workers employed to sustain these 'landscapes of privilege (Conway & Leonard 2014). The next section further examines the making of British 'expatriate' subjectivities through these embodied interactions.

Encounters with Low-income Migrants

...when people of apparently different races and classes find themselves in slow, crowded elevators [...] intimacy reveals itself to be a relation associated with tacit fantasies, tacit rules, and tacit obligations to remain unproblematic. We notice it when something about it takes on a charge, so that the intimacy becomes something else, an 'issue' – something that requires analytic eloquence (Berlant 2000: 7).

Proximate encounters of physical and verbal exchange between British migrants and low-paid migrant workers employed in the service, retail and hospitality sector are a routine part of everyday life in Dubai. In this section, I focus on the way in which encounters with low-paid migrant workers from south Asia were discussed and enacted by interviewees (in Chapter 6 there is further reflection on cross-cultural interactions both with middle-class migrants in nightclubs and with sex workers). A critical analysis of how such encounters are framed by British migrants is revealing of the ways in which such accounts are tied up with their understanding of themselves as British migrants and the ongoing construction of 'expatriate' subjectivities in the 'postcolonial city'.

One of the most frequent and mundane encounters British migrants have with those of another nationality takes place within the home with their employees: migrant domestic workers. The sharing of domestic space leads to an intimate knowledge of the everyday lives, bodily routines and family relations of the employing household. Indeed, the intimacy of these interactions is countered by distancing mechanisms, such as the enforcement of the wearing of uniforms or restrictions on household spaces (Constable 2008). Britons among my interviewees varied considerably in their attitudes to the employment of domestic workers. Not all British households employed domestic workers: some households instead paid an hourly rate to a cleaning company to provide daily or weekly cleaning. Often this was related to the size of the household and biographical variations in previous work and migration patterns. For instance, a couple previously living in the UK, with a dual income and without children, and used to sharing domestic chores or already employing a cleaner, was more likely to consider a live-in domestic worker unnecessary, as were single British migrants who lived alone in apartments or shared villas. However, it was relatively common for British households with children to have a 'live-in' domestic worker whom they called a 'maid'. For those with 'expatriate' experience in other countries this may even have been a reduction in the staff they were used to. As Eve (forties, accompanying her husband) explained: 'in Pakistan or Bangladesh, it is common to have so many people around the house you don't know who half of them are' (interview).

It generally falls to British women to manage any domestic workers employed by a household (see Chapter 7 on the partial re-domestication of British women through migration). I witnessed some resistance to the role of employer while, simultaneously, it was a role through which 'expatriate' subjectivities were reproduced. Showing me around her home, Claire explained that she had attempted to be on first-name terms with her domestic worker, while at the same time suggesting that subordination is culturally determined and fixed:

> You have to get your mind around it. In your mind it still smacks of slavery, you feel embarrassed, but if you're paying them well, it's all you can do, because they don't have conditions or rights. I couldn't be doing with fish-fingers and children's dinners and all that. I have asked her to call me Claire, but to my face she calls me Madam. It's her culture. She thinks it should be that way, which is fine. Of course the Sri Lankans are very efficient. Once you invest a bit of time in them they are very good. I liked my last maid and she was Sri Lankan so, even though this girl has no experience, I am tempted. Then I can tell her how I like things done anyway. It is nice to have someone who comes in and magically knows what needs doing and you get back and 'Ta-Da!' I spoke to someone and she said the going rate for live-in was nine hundred Dirhams a month [approximately £130] as she's got no experience (field notes).

As Nicole Constable (2008) has noted from her extensive research with domestic workers, nationality is often used as a simple way of distinguishing between the likely characters and cleaning skills of maids, reflecting well-established employment hierarchies. The construction of the Asian domestic workers as 'unskilled', 'immature', and 'inexperienced', all evident in Claire's comments, while not as a blatant as the colloquial 'tea boy' and 'house girl' sometimes used by nationals (Kanna 2010), were the kinds of framings that helped the British migrants who employed live-in domestic workers to justify low wages in return for long hours, as well as condescending employer–employee relations. Domestic workers are expected to be continually present in the home, yet there is some ambiguity over whether their labour is supposed to be observed. Lucy (mid-forties, accompanying spouse) complained, 'This particular one, I'm sure she used to hear the car pull up and come in and pretend to be busy because nothing was ever done right' (Lucy, mid-forties, trailing spouse). Similarly, Malcolm (teacher) commented: 'How can he really get into it if he's not wearing overalls? He's so slow as well, you know if I was cleaning this house I could do a much better job of it in half the time. I don't think he uses the chemicals we buy. They just splash a bit of water around and think that's clean' (field notes).

Encounters with taxi drivers are another example of where British migrants come into close contact with low-income migrants, though this time the relationship with any particular individual is not sustained. Taxis were used frequently in Dubai by the young single British migrants I lived among, some of whom did not have their own car, while others needed to observe the zero-tolerance law with respect to drink-driving. Kendall (2012) notes that the climate, empty sandy lots without pavements, and the huge highways, discourage people from walking in Dubai unless they are forced to by poverty and lack of access to public transport, like the construction workers he photographed walking long distances from their labour camps. Unlike such labourers, British migrants can both afford and access buses, yet to walk to a bus stop and endure an hour's journey, rather than to order or hail a taxi and be driven direct, was unthinkable to those of us without our own cars, even in winter. Although taxis cost much less per journey than they do in the UK, their use is also part of the production of 'expatriate' subjectivities, a spatial claim of ownership through a relative ease and speed of mobility. Our migrant topographies were therefore marked by our distinction from low-income bus users. Taxis are most frequently driven by migrants from south Asia (although also some poorer Middle Eastern countries); male migrants from Sri Lanka and Pakistan in particular. Since demand was rising rapidly in the early 2000s, taxi drivers at this time were often very recent migrants. The city's roads were also undergoing constant change with the expansion of the suburbs and new construction projects. It was hardly surprising, therefore, that taxi drivers might have had difficulty navigating Dubai, and the taxi frequently became a site of conflict, especially when British residents had consumed excessive amounts of alcohol. During one such journey, one of my English compatriots called the driver an 'idiot', swore many times, and deducted money from the metered fare since the driver made a wrong turn. When I questioned Malcolm, he defended his actions by positioning both the taxi driver and himself as migrant workers, attempting to obscure any racialised or classed status hierarchy: 'But it's all relative! In their own countries they have nothing, so they're much better off. Some of them are really happy…they're happy with their lot' (field notes). Interestingly, he also felt marginalised by the changing status of British migrants in Dubai: 'we're not living the expat lifestyle that people at home would imagine. I wouldn't say we're carving out an existence, but we're not affluent [either]' (field notes).

In interview settings, however, Britons frequently recognised and articulated a sense of privilege in accordance with their position in Dubai's social hierarchy, even if it was a counterpoint to their everyday interactions. Ed (late twenties, recruitment consultant) reflected:

> Things would actually turn my stomach when I first got here, but now I find I'm the first one to shout at the cabby when he doesn't know where he's going, but it is more huffing and puffing than being outright offensive (interview).

Indeed, interview accounts of everyday life were sometimes deeply reflexive of the inequalities that structured Dubai society. An interview extract from Paul, a journalist in his mid-twenties, provides an example:

> To me it's a Pinocchio scene, a fantasy land where kids are allowed to do what they dream about: being corrupted. There are lots of people who are eleven-year-old boys essentially but being allowed to do shit that they would never normally be allowed to do, and it turns them into donkeys to a certain extent, in terms of racist, um, and they're likely to be cheating on their wives, and they drink-drive, and they drink too much, and so on. And I'm totally in with that as well, after only six months here. I find myself becoming deeply colonialist, immoral, racist. I find myself becoming so racist since coming out here. You just look around you, how this city is set up for you, and you think: 'I've fucking made it!' I wonder what happens to expats when they go back. I wonder how on earth I'll feel going back, if I'll suddenly feel absolutely inadequate and sort of weird and suddenly stripped of a lot of status and prestige. I've found that to be someone a bit special, you need to be a foreigner (interview).

There is something about the space of the home or taxi being considered a 'private' space, in spite of it being also a work space, that seemed to heighten the likelihood of British migrants behaving in such a way. While these are intimate spaces in the sense of bringing people physically closely together and offering a possibility for communication across cultural boundaries (see Schlote 2014: 40), research observations suggested that they were rarely the site of the development of intimacy in terms of an empathetic relationship. Fortunately, British migrants were usually polite in their encounters with low-income migrant workers in the more highly regulated privatised spaces of hotel bars/restaurants, shopping malls and hotels. Furthermore, when British migrants enter public spaces dominated by low-income migrants at leisure themselves, their own vulnerabilities and the instabilities of whiteness may even be exposed (see earlier section and Chapter 5).

Conclusion

I cannot do justice in the space of a single chapter to the variation that exists in terms of Britons' interactions with Emirati nationals and low-income migrant workers. As such, while I have tried to evidence moments of resistance to the automatic reproduction of a privileged 'expatriate' subjectivity in each section, it is also necessary to restate here that there is a diversity in attitudes, experiences and subjectivities beyond the dominant narratives detailed here. Nevertheless, certain observations were so frequent and widespread that it is possible to draw tentative conclusions. Firstly, meaningful social interactions with Emirati nationals were extremely rare beyond the workplace and the high status of Emirati citizens provided few opportunities for cross-cultural friendships or romantic relationships. As a result, British migrants tended to understand the Emirates though touristic, bureaucratic and legal encounters and construct British 'expatriate' subjectivities through binary discourses of self/other or British/'Arab'. The relational dimension of these constructions is evident from the accounts in which Britons compare UK and Dubai norms. Encounters with low-income migrant workers, meanwhile, are shaped by employer–employee relations and also shot through with notions of culture, race and ethnicity that allow us to trace colonial articulations of difference in the postcolonial city.

US scholarship has portrayed a society in which public intimacies are increasingly supplanted by domestic or privatised commercial forms, such as gated residential communities and shopping malls (Wilson 2012). In this chapter, I have demonstrated how the exclusive spatialities of urban leisure contribute to the racialised and classed segregation of migrants and the regulation of cross-cultural proximities in Dubai. Existing scholarship shows us that low-income and middle-income Indian nationals resident in Dubai live out much of their everyday social lives within walkable inner-city neighbourhoods (Elsheshtawy 2010; Vora 2013). The lowest income migrants, housed within the labour camps beyond the city or in industrial areas, are even further removed from 'new Dubai'. In contrast, British migrants are relatively privileged within the labour market, social hierarchy and cultural politics that operate within 'postcolonial' Dubai, enacting affluent 'expatriate' subjectivities racialised through whiteness. As such, they not only have access to these privatised residential and leisure environments, but rely upon such spaces in the reproduction of their British subjectivities in contrast to the status of migrants who are excluded from notions of 'consumer citizenship' (Vora 2013) by poverty and private security guards. The geographies of privatisation in Dubai allow for a separation along lines of affluence and nationality, creating a controlled mode of intimacy.

In the case of British migrants, the types of intimate personal relationships I am exploring in subsequent chapters of this book – friendships, couple and family relations – are nearly always restricted to fellow nationals, sometimes extending to other European or US and Australian citizens. In this chapter, I have tried to detail the social setting of Britons' residence and relationships, especially in terms of segregation based on race and class, in order to explain why national boundaries might be so strong in Dubai and how it is that Britons consider this 'natural'. The lack of meaningful interactions with Emirati nationals and low-income migrant workers, the segregation and privatisation of urban space, and the imaginaries of travel and whiteness that accompany Britons are all contributing factors which, together, shape the everyday lives of British migrants in Dubai and contribute to the making of 'expatriate' subjectivities. Although there is some diversity and resistance in evidence, it appears to be difficult to live alternative lifestyles against the social norms regulating urban intimacies. In Chapter 5, I examine the communities and friendships emerging in Dubai, mostly among Britons, shaped by the migration status, privilege and spatialisation of British 'expatriate' subjectivities I have described in this chapter.

Chapter Five
'Community', Clubs and Friendship

In comparison with the sexual/romantic and familial relationships discussed in later chapters, much more is known about the 'community' lives of British migrants from existing empirical studies in other countries, including Hong Kong (Knowles & Harper 2009; Leonard 2010), Singapore (Beaverstock 2011), and southern Spain (O'Reilly 2000). In Scott's (2007) study of British migrants in Paris, it is possible to discern a huge range of different kinds of Voluntary Community Organisation (VCOs) related to: business, charitable causes, religion, education, culture (e.g. St George's Society) and clubs and societies (e.g. organised around alumni, mothers/women, theatre, youth and sport). As this chapter will explore in more detail, a huge range of VCOs were also evident in Dubai, where they fulfilled a number of different functions within the community. As we might expect from Beaverstock's (2005: 246) assertion that highly skilled transnational managerial elites 'embody the cross-border circulation of knowledge, skills and intelligence', for some British migrants the British Business Group affiliated with the British Council was a significant site of networking. Likewise, non-working women's networks were often generated around British and international schools or coffee morning groups, such as the Dubai Adventure Mums. However, I also identified an increasing number of Britons for whom these more formalised institutional settings were not the focus of their social lives (see also Scott 2007) and who may never have set foot in the British Club. This chapter, therefore, encompasses discussion of the sociality of British migrants both in terms of VCOs and the friendships beyond, including in shared households (with Chapter 6 extending this by tracing nights out in the city). Further, the chapter

Transnational Geographies of the Heart: Intimate Subjectivities in a Globalising City, First Edition. Katie Walsh.
© 2018 John Wiley & Sons Ltd. Published 2018 by John Wiley & Sons Ltd.

attempts to extend our understanding of British migrants' social lives by going beyond a description of their social networks to explore the emotional geographies of the negotiation of friendship. Social scientists consider friendship to be 'an increasingly important form of social glue' as 'more people are culturally determined by their friends until much later in life' (Pahl 2000: 1). Through attention to British migrants' narratives and practices of friendship, therefore, I demonstrate how the mobilities, temporalities and translocalities of 'expatriate' migration in Dubai shape the textures of intimacy at play in their everyday lives. In doing so, I extend the existing understanding of institutional networks in 'expatriate communities' put forward by Scott (2006, 2007).

It is understandable that existing studies of British migrants have highlighted their institutional social lives, rather than the more emotional accounts of their friendships that I draw attention to here. Wider research on migration and transnationalism has tended to assert the significance of VCOs in theories of social capital and integration, while research on the highly skilled from a globalisation perspective is dominated by an interest in their working lives and the associated circulation and transfer of knowledge. Institutional lives are seemingly, too, more readily accessible to social science researchers than other kinds of relationships: arguably, they are the more public dimension of our personal lives. Logistically, too, institutions are often the entry-point for ethnographic work and interviewing strategies, with their chairpersons acting as important gatekeepers to the wider 'expatriate community', so it makes sense to understand how institutions are functioning in the lives of their members. Nevertheless, the institutions respond to, and help to produce, gendered, racialised and classed textures of intimacy, so access is not always straightforward: the positionality of the researcher informs the kind of institutions from which we might be excluded or those we inhabit more freely.

It is important, however, not to draw a discrete boundary around these institutional lives as the enactment of a particular kind of social relationship only. While some forms of institutional social life may not be especially intimate, the institutional lives of Britons in Dubai are entangled with their more intimate relationships; for example, couples and families may socialise together within wider co-national or 'expatriate' networks. Club spaces also have an interesting connection with relationships based on friendship. They may both stand in for 'real' friendships and be the starting point for deeper friendships to evolve. Nonetheless, in this chapter, I echo Conradson and Latham (2005b) in urging research on friendship specifically. If we focus only on clubs and institutions, then we miss out on a significant part of migrants' sociality. Likewise, while there are shared and overlapping meanings to a degree with the related concepts of love and family, friendships are not captured by these terms with which migration studies is seemingly more comfortable.

For Bunnell et al. (2012: 491), geography 'is important in the making, maintenance and dissolution of friendships, as well as in the types of friends that are important within particular space-time settings'. Material spaces, they remind us, 'constitute the key technologies of friendships' and 'friendships are also productive of lived spatialities' (ibid.: 491). In spite of their privilege, British migrants are a minority in Dubai and mostly seek out social spaces in which they expect to find people similar to them and be able to mobilise a sense of 'community'. The homogeneity of many friendship groups is more widely noted by geographers. Gill Valentine's (1993) early article on friendships among lesbians showed that the heterosexuality of everyday space led to them choosing gay social spaces when seeking friendship, such as gay nightclubs, social events, drama and sports clubs, support and political groups. Similarly, Bowlby (2011: 608) notes that 'the existence of friendships between people is often an indicator that they share salient social characteristics, especially class, ethnicity or gender', such that 'one function of friendship can be to provide various forms of emotional and practical support between those with similar power and status'. As such, I start this chapter with attention to collective social spaces and groups based largely on nationality and 'expatriate' identities, in which performances of belonging are also classed, ethnicised and gendered. These groups are often the context within which friendships are established and partly enacted, so recognising how the groups establish themselves through boundaries which are simultaneously inclusive and exclusive can help to maintain a critical approach to friendship. Subsequently, the analysis turns to how friendships *feel* different in Dubai, moving the discussion into a more emotional account of friendship in which some of these broader cultural politics of the city are absent in participants' narratives.

'Britishness' and Club Life

Towards the end of my fieldwork period, in early 2004, the British Business Group for Dubai and the Northern Emirates (BBG) held a 'Great British Day' at a five-star resort hotel, Le Meridien Mina Seyahi. Business groups oriented towards a single nationality are common in Dubai. The purpose of the BBG, it was explained to me by a member, was to promote and strengthen bilateral trade between the UAE and Britain, 'while having a little fun along the way' (interview). There had been seven previous 'Great British Days' held in Dubai, but this was to be the first after a gap of a couple of years. The BBG committee had selected 'the medieval period' to theme the event and the key signifier of this was a display of jousting, with knights in full costume, and the fence to hold back the crowds was adorned with shields decorated by children attending British schools. Other activities were reminiscent of a country fete; for example, there were stalls with games, such as a tombola, apple-bobbing, and face-painting. Entertainers were dressed as clowns and there were bouncy castles. The hotel catering staff, migrant workers from south and south-east Asia, provided guests with an all-day buffet of familiar or symbolic cuisine for the British palate: shepherd's pie, fish and chips, and haggis were among the extensive offerings that covered vast tables in the hospitality area of the lawned grounds. This, it seemed, was a day for Britons in Dubai to gather together in an overt performance of Britishness and collective heritage and, in doing so, mobilise and construct a British 'community'. Interestingly, however, this event was not taken seriously by most British migrants I knew; rather, they engaged with the theme of national identification it engendered through a sense of ironic playfulness. In a sense, then, it obscures the significance given to the more everyday sense of Britishness that I described in Chapter 4 as something experienced by Britons when they are unsettled by bureaucratic or public encounters. Nonetheless, the BBG's production of this spectacle of 'Great British Day' would not have been possible without the existence and significance of an infrastructure of British 'community' institutions and networks in the everyday lives of Dubai's residents. Building on Chapter 4, then, this section contributes to an understanding of the making of a British 'expatriate' subjectivity through the clubs and institutions that structure their sociality in Dubai.

In studies of British migrants in postcolonial global cities in Asia, the significance of the permanently sited British club as an institutional feature of community and social life has been clearly demonstrated (Beaverstock 2011; Knowles & Douglas 2009). Often informally referred to in conversation as 'the British club', Dubai Country Club opened in March 1971. It was located some way out in the desert, in Ras Al Khor, on land gifted by Sheikh Rashid who also ordered the building of a road to the club (Croucher 2011). It was the first golf course in the region and had one of the only bars serving alcohol when it opened, making it an enormously popular club with British migrants in the 1980s and 1990s. By the time the club closed in 2007, to make way for the development of the Meydan racecourse, membership had declined to 850 (Croucher 2011). There is no doubt that this was in part due to the impending uncertainty about its future but, even by late 2002 when I was starting my fieldwork, the vast majority of British migrants I met – families and single migrants alike – had no involvement at all with the Dubai Country Club. By then there was a wider range of membership choices in clubs, as well as a huge explosion of pubs and bars around which a non-institutional social life could be conducted (see Chapter 6). However, my own access to the Dubai Country Club was limited not only by my not being able to afford membership and my lack of association with other members, but also the distance from the main city that made routine visits prohibitive without car ownership (something that would have impacted on the many younger Britons who relied upon taxis). I visited on only two occasions: firstly, by prior arrangement with the management for interview purposes where I was also introduced to a women's daytime social group, and, secondly, to attend a Christmas Eve carols event. Festive celebrations such as carol singing are typical of the kind of events that clubs organise for their expatriate members, with more routine quiz nights and sporting competitions also in evidence to supplement the provision of facilities (see also Beaverstock 2011). My lack of participation in what was essentially 'the British club' was a disappointment to me, but fortunately is less problematic for examining British migrants' intimate subjectivities. Not only is the role of club spaces in the production of white expatriate subjectivities already well understood (e.g. Knowles & Harper 2009; Beaverstock 2011), but their relative lack of importance for so many British migrants in Dubai at the time was evident.

Of more significance in structuring the social lives of a wider section of the British community were nationality based institutions that did not have a locatable site within the city but instead organised events in different hotels, notably the St George's Society and Dubai Caledonian Society. These groups tended to mobilise those who had lived in Dubai for a long time or those who were career expatriates, for whom Dubai was one posting among many. One British member referred to such societies as 'useful, particularly as a way of meeting like-minded people' (field notes). He explained that he got involved in the society 'believing in the importance of participation in the wider community', but this was not a vision of 'community' as inclusive of all the residents of his city or proximate neighbours. Rather, as became apparent, this was an 'imagined community' (Anderson, cited in Valentine 2001: 104–105) based on nationality:

> You do mix with your own. It's easier. Mixing with so many nationalities at work, when you're out, you want to relax. It's a common way of trying to build up a community because you don't have that abroad (interview).

Thereby, this committee member mobilised the idea of seeking out a national community, without a residential base in a particular neighbourhood, as a natural response to difference and diversity.

Other social groups were connected by shared membership of a particular space or 'club', but one based on affluence or broader 'expatriate' members rather than nationality. Examples would include golf clubs, hotel beach clubs, health clubs or those connected with larger residential compounds or apartment buildings. In these private spaces, members use the sports and fitness facilities provided or participate in classes or group activities at regular times. However, they may also choose to spend other leisure time in the on-site, partly subsidised, bars and restaurants, usually in couple, family, or friendship groups. For example, Susan and Tom were members of a golf club where Tom played golf with his male colleagues, Susan sometimes took her grandson after school for a milkshake when he asked to go to 'Grandad's club', and the couple met friends for a meal or to play Bingo on a Tuesday night. They were kind enough to invite me along to a quiz evening and I witnessed their familiarity and ease with the space and its other members, many of whom they talked with over the course of the evening.

Another example is the Dubai Offshore Sailing Club that was established in 1974. While it is now unrecognisable, in the early 2000s this club was still quite small, with low-key facilities. I was invited along by Janet to accompany her and her children for 'supper' after school one evening, her husband being away on business. Clubs like these are not simply spaces for expatriates to socialise, as the secretary of the club explained on another occasion: 'the club is strictly a sailing club, so there must be at least one "sailor" for each family's membership' (interview) and sailing lessons were made available for children and those without experience. However, as my visit to the golf club also hinted, there is more to the sociality of these spaces than participation in particular sports. This sailing club organised evenings for new members to meet, a playgroup, a weekly quiz night, sailing evenings themed around different classes of boat to meet those with common interests, a cadet club for children, and regattas each weekend. When we arrived at the sailing club, we entered past a notice board and Jane, Janet's daughter, proudly pointed out the pictures of boats for sale that were like the ones she had used for her lessons. We went to the small bar to order food, but Jane ran behind it and helped herself to a glass of ice, informing me: 'this is what I do'. Later on, we ate pies and burgers, served with chips and gravy. Janet's two children played outside with their friends, reappearing throughout the evening only when they were thirsty or wanted ice-creams. Two of Janet's friends arrived with their husbands and children, and a quiz was held in the bar. There was amicable squabbling over half-points in the quiz which, in the end, was won by a noisy group of other regulars. The next morning was Friday, the weekend, so I accompanied Janet to the sailing club again. After her cadet club lesson, Jane sat on a stool at the bar reading, while her brother Ben went swimming. Later they used the playground together and then it was time for chicken nuggets for lunch. As this suggests, golf, sailing and even hotel beach clubs provided a members-only privatised space of leisure for couples, families and single British migrants to hang out with other 'expatriates': a homely space characterised by a sense of ownership and familiarity. Inhabiting such spaces, British migrants reproduce their 'expatriate' subjectivities as racialised and classed subjectivities. In smaller Gulf cities, such as Manama, Muscat and Doha, where the numbers of British nationals are fewer, there tend to be much stronger, more visible and explicitly labelled British VCOs, but in Dubai there are more clubs and societies and many of these are Western and Anglophone 'expatriate' communities rather than exclusively British. Furthermore, the British nationals in Dubai are increasingly diverse among themselves and in the next section I explore how this is shaping their sense of 'community'.

Heterogeneity within the British 'Community'

In their study of British expatriates in Hong Kong, Willis and Yeoh (2002: 559) remind us that 'communities of transnationals' are not given but, rather, 'constructed and develop through the actions of individuals and thus may seem to be exclusionary to those who do not fit into the widely accepted and perpetuated mould of an expatriate with particular gender and nationality characteristics.' Expatriate communities are also subject to change over time; Scott (2006, 2007), for example, identified the changing social and community morphology of British migrants in Paris. The increasing diversity in motivations for migration, occupational groups, and lifestyle types among the Britons in Paris, he argued, is reflected in the change in how the community is functioning, 'from one centred on "traditional" expatriate enclaves, to one involving more diverse VCOs, as well as increasingly tribal-ephemeral networks and growing transnational linkages' (Scott 2007: 655).

Older interviewees in Scott's (2007) research remembered a smaller and more cohesive community of privileged expatriates, in contrast to the diverse group he encountered during his fieldwork. Scott (2007: 1112) identified a typology of British residents of Paris based on distinctions in their motives for migration, commitment to the host city, family status and lifestyles. Among the professional career-oriented migrants, there were family migrants whose move tends to be temporary (between 1 and 5 years) or established 'lifers' whose adult children had left Paris, but who had helped to create the original institutional fabric of the British community in terms of its VCOs. However, among the professional career-oriented Britons in Paris were also young singles, among whom international employment is seen as prestigious and working identities are omnipotent. In contrast to the family migrants, they tended to socialise informally after work in 'tribal-ephemeral networks' based on pubs and bars (Scott 2007). Scott (2006) further identified bohemians, typified by 'a cultural, intellectual and/or artistic engagement', living in modest accommodation, and engaging with an international community, as well as graduate lifestyle migrants whose career aspirations are secondary to their desire to travel and the cultural opportunities of world cities. As I discuss later in this chapter, in Dubai it is also possible to observe a group of non-graduate young Britons employed in the service sector in hospitality and tourism, for whom lifestyle and economic opportunities appeal. The final sub-group of the British community in Paris discussed by Scott (2006) are those he terms 'mixed-relationship' migrants, in other words, those who are in a relationship with a French national and are, therefore, much more embedded into French social networks. In Dubai, this is a tiny minority of Britons, since only a very few British women are married to Emirati nationals (see Chapter 4).

For Scott (2007) this shift from a relatively cohesive community to diverse ways of socialising as a Briton reflects the expansion and fragmentation of the British middle classes back in the UK, itself associated with increasing access to higher education, growing affluence and the importance of consumption practices in performing varied middle-class identities. He goes further, however, in suggesting:

> it is plausible to see international migration as a particular mobility strategy, employed consciously or otherwise, that leads to the appropriation of social, cultural and economic capital [Bourdieu 1984]…important in middle class reproduction and internal distinction (Scott 2007: 1109).

While not identical to the range of types of migrants in Paris or other global cities, Dubai also had, by the early-to-mid-2000s, an increasing diversity evident among the British migrants in residence there. As a result, performances of Britishness were heterogeneous and contested, sometimes bringing the very construction of a 'British community' into question. This section highlights social cleavages based upon class, ethnicity, religion and age/marital/family status. I argue that the mobilisation of 'like- mindedness' and 'similarity' observed in the narratives of British VCO members in the previous section is an appeal to a very specific sense of Britishness in which Englishness dominates and, therefore, the construction of collective belonging is classed, racialised and heteronormative.

Britons of a minority ethnic heritage represent an increasing proportion of the British residents in Dubai and they frequently recognised this in conversation. This contrasts with the way in which white Britons assumed that their own sense of nationalised belonging is unmarked by race and ethnicity. For example, I met Hanif (forties, financial services) when he was participating in a St George's Day Ball, but in a later interview, he explained:

> Dubai society is cosmopolitan, but the actual mixing up doesn't happen much. The Brits here are very focused on their community, Dubai encourages it. Expatriates out here might preach equality at home, and we've got laws against discrimination, but it's all forgotten out here. I don't mix with the English British, it's very hard. I've been British since birth so I'm British, but not English, do you see what I mean? Still it's nice to talk about home and I have met another British Indian and we get on very well and go out drinking and all that (interview).

Increasing class diversity was also evident, related not only to the socio-economic background of Britons before migration, but also to their occupation, seniority in respect to their career trajectory, and skill sector. At the 'Great British Day' event I described earlier, for instance, Sally, a long-settled British entrepreneur commented:

> You see the best and worst of the Brits on days like this. We had the family from hell sat next to us earlier, you know, drinking beer, eating chips and the kids having tantrums! There are some right yobs living here nowadays (field notes).

Furthermore, there is something of a classed social division in the British 'community' that has emerged between married couples (often with children) and single migrants. The former frequently have 'expatriate packages' based on the highly skilled, highly paid employment of the lead-migrant (usually male). In contrast, the increasing numbers of single male and female migrants are usually employed in lower skilled positions than their married counterparts because they are younger and at the beginning of their career. In addition, there are some single migrants who are employed in non-graduate roles, for example as fitness instructors or hotel/bar management. This became evident during my fieldwork in Dubai where my own positionality (as a young, single, unmarried, low-income woman) was an obstacle to fully accessing many British institutions. In my field diary I reflected on the impact of my positionality on the embodied experience of fieldwork. For example, I received an open invitation sent by email to members of the Dubai and Sharjah Women's Guild to attend a church service for Remembrance Sunday. Once there, however, I was given a ticket to a reception in the garden of the Consul for the British Embassy for 'Drinks on the Lawn'. There were scouting leaders, church leaders, the flag-bears and their parents, members of the Women's Guild and other British societies, such as the Caledonian Society, as well as 'the oldest, longest, British resident of Dubai', as one person whispered to me. I invited one of my key interlocutors, Malcolm, and, as we queued up for drinks, he joked: 'Let's get lashed [drunk] on the embassy', reflecting the unusualness of our situation. We had trouble mingling on the lawn for small talk – it was probably one of the most awkward events I attended during my fieldwork. We were looked at, but no one approached us, so we decided to separate to try and 'infiltrate'. Malcolm headed confidently into the middle of a large male group. He reported back afterwards that he had met and talked with 'probably the highest paid man in Emirates' and the 'diplomatic elite'. I recognised a couple of Women's Guild members, so I attempted some small talk. I heard myself responding to their questions in a proper accent and refining my mannerisms. Noting my discomfort, perhaps, some of the women patronised me. Deirdre introduced me to everyone connected with her church: 'This is Kate, Katie… (she turns to me) Which is it? (hardly giving me a chance to speak)… not a Katherine then? She's studying expatriates! Yes, us! Do you remember that documentary…' Without the common focus of our monthly meetings in the church hall, I felt out-of-place among the older women, themselves seemingly so much more at ease in the British compound, and it was a relief to leave. As we did so, Malcolm and I joked about going 'back to Satwaaa', the low-income predominantly south Asian neighbourhood we lived in, 'to hang with the massive'. As a teacher himself, Malcolm found this opportunity to interact with the highly skilled British elite highly unusual and I heard him recounting the episode to his friends later that evening.

Belonging, Leisure, and the Significance of 'Play'

The vast majority of the VCOs that British migrants were engaged with were activity-based groups, including societies based on particular hobbies or sports. For example, some important groups among the British community at the time included: Dubai International Arts Centre, the Dubai Natural History Society, the Dubai Chamber Choir, the Dubai Harmony Choir, the Barbie Hash House Harriers, the Desert Beats salsa group, and many more. These were mixed-nationality groups, though dominated by white Western 'expatriate' migrants. I have touched on the significance of these groups in terms of those which provided members-only 'expatriate' spaces, such as the golf and sailing club. However, to further understand the significance of these groups, I turn now to studies of leisure in which geographers (e.g. Crouch 1999) have argued that leisure is a primary factor in many people's social identity. Understanding leisure as practice, as Crouch (1999) encourages us to do, enables us to explore the 'feeling of doing' as being about fleeting, sensual practices that create a bodily knowledge. Furthermore, Nielson (1999) suggests that in contemporary societies, leisure activities bring about a sense of belonging: creating a relational context to be part of and producing knowledge of who we are by 'doing' (see also Crouch 1999). Nielson argues that identity is gained by moving and living in a space that you appropriate by your body and not by your thinking, so leisure provides the possibility of making a sense of home through a multiple sensory range of experiences 'grounding' belonging. Arguably, participating in a hobby puts the expatriate in a 'habitual space', where the actions are familiar and intelligible to them, and where they do not need to try and make sense of what is going on (Fortier 2000).

Participation in these kinds of activity groups appeared to bring an enormous sense of well-being to British migrants. I have chosen three, seemingly disparate, examples from my fieldwork that demonstrate the affective relations of the leisure activities engaged with through membership of VCOs:

> The Dubai International Arts Centre has a membership of over 1000 people, incorporating more than fifty different nationalities, who come together to join workshops and classes to complete art projects ranging from pottery to mosaics, painting or photography. During the afternoon class, a hush fell over the studio as everyone becomes absorbed in their painting. The teacher Trey wanders around the room quietly, giving advice, referencing to colour and movement. Cynthia gasps, startled by his approach to her easel – she is lost in concentration, delicately adding highlighting colours to the waves of her seascape. There is a pause for a coffee break, but the talk remains focused on the paintings. (field notes).

> I'm at the Exiles Rugby club where two expatriate teams are playing a league match. The evening is exhaustingly humid, even just watching. The men are caught up in the game: yelling to each other, running, passing, catching, throwing, kicking, and skidding on the grass. The tension rises in the second half, until a try takes one side clearly into the lead. Afterwards at the bar, the talk initially is all about the play on the field (field notes).

> The Dubai Equestrian Centre runs horseriding classes for all abilities and can also provide individual lessons and stabling facilities. The equestrian centre is bright with floodlighting. I can hear the sound of hooves on the cobbles as horses are led between paddock and stable. It is busy: at five in the evening, there are a lot of lessons going on. I wait with Tanya for the children who precede us to finish their lesson. We're excited. It's so nice to be here at the end of a stressful week. Tanya confides 'I wake up thinking, "Great, it's riding today." I always look forward to it all day.' We ride for 45 minutes, forgetting work, people, everything, listening only to the instructor as she

calls instructions: 'good, keep going…ok, now trot…slow the trot down, slow it down…keep your heels down…good, now halt, slowly…bring your reins in…'. Afterwards we're elated with our progress, the endorphins are flowing, and we stagger happy and tired towards the car, reliving the dramas of the lesson (field notes).

While VCOs are highlighted already in the literatures on migration as a way of meeting people and making friends with similar interests, these field notes suggest that there is also something additional to be understood about these kinds of playful leisure activities and their role in making the expatriate feel at home. The concentration involved in making a 'good pass' to a team mate during a rugby match, mixing the exact shade of blue desired for an acrylic painting in an art class, or controlling your body to communicate with a horse, is distracting. There may be little conversation, or it may centre on the activity, focusing the mind away from the outside environment, particularly towards the actions of the body or the perceptions of the senses. These activity clubs are also about meeting oneself: in concentrated relations, in practices of creativity, in skilful but extra-cognitive actions of 'play' (Nielson 1999).

Coffee Mornings or Adventures for 'Expat Mums'

Among British couples, it is usually the male partner who would be understood as the 'lead' migrant in that his employment generates the relocation. As such, men usually find themselves participating immediately after arrival in Dubai in professional collegial networks that are initiated in, but extend beyond, the workplace. Indeed, the significance of 'after-work drinks' for expatriate socialising has frequently been identified (e.g. Beaverstock 2005; Knowles & Harper 2009; Scott 2006). In contrast, British women who accompany their husbands in relocating to Dubai must instead forge their own networks. This is seen to be vital to the success of expatriation; as an orientation specialist in Dubai commented: 'the men are always straight into work but it worries them if the wives don't have anything to do…they get it in the neck at night!' (interview). Women with younger children often made friends through school networks, but this may not be sufficient contact to develop a friendship and was often described to me as intimidating by new residents. As a result, several social groups in Dubai catered only for women, organising events during the daytime: The Dubai Adventure Mums; breast-feeding and post-natal support group Mother-to-Mother; The Dubai and Sharjah Women's Guild; and Dubai International Women's Club (begun in 1967 as the Petroleum Wives Club). In spite of my status in Dubai as a single woman without children, these were the institutional groups that were more accessible to me as a result of my gender. I carried out ethnographic research in both the Dubai and Sharjah Women's Guild, as previously mentioned, and the Dubai Adventure Mums, which I discuss further here.

The Dubai Adventure Mums was a non-profit organisation started in 2001 by a small group of women. Many of the Dubai Adventure Mums were British, but there were 34 different nationalities among members. The club worked mainly by word-of-mouth: women were often invited initially, brought along to one of the monthly coffee mornings by someone else. Although women were then able to join more officially by paying a small membership fee, there was no pressure to attend on a routine basis. The group also stopped during the summer when most of the mums took their children 'home' (with or without husbands) for the summer holidays. During my participation over the course of a year (February 2003 to February 2004), the Dubai Adventure Mums met twice monthly, once for coffee only and once for a 'challenge'. For instance, I attended a bowling morning where my team called itself 'Blighty's Bowling Babes', an indoor abseiling challenge at the Wafi Centre, and an introduction to golf at the Fairmont Golf Course. These activities were central to the aims of the group. The committee explained their rationale for meeting:

> There's no point in coming to a country and staying in having coffee. So, it's all about getting out and exploring the country you're in. Then, if mum likes it, she'll get her family out. You have to leave the country with a treasure trove of memories, of experiences. Friendships develop straight away, it's really good… the activities allow them to be silly and break down their own social barriers (interview).

The quad-biking morning on Big Red (a sand dune) challenge was introduced at a coffee morning with a similar emphasis: 'I believe a happy mum makes a happy family and a happy home…come along and have a go, there will full instructions and it'll be a lot of fun. You'll be able to tell your husbands something exciting over dinner' (field notes).

It is intriguing that this organisation framed its adventures against the stereotypical activity of the coffee morning. Anne-Meike Fechter (2007: 106) notes that coffee mornings are 'one of the most recognisable institutions of expatriate women's lives' alongside charity work and handicrafts. In her ethnographic research, coffee mornings held in Jakarta's finest hotel ballrooms emerge as a way of performing expatriate identities, spatially removed from Indonesian culture and Indonesians themselves (apart from waiters). Organised by nationality, British women's coffee mornings involve loose leaf teas and freshly baked muffins that can be read 'not simply as a re-enactment of "being British in Britain",' where ordinary biscuits and teabags would suffice, but 'an exaggerated form of Britishness' (Fechter 2007: 106). In contrast, British women in Dubai were seemingly keen to dissociate from such performances: they might have coffee, but they also had adventures. The activities brought a language of empowerment to the group during their duration as well:

> At the Wafi Centre, next to their indoor climbing wall, we're putting on our harnesses and laughing at how they make our bums look unattractive! The two

instructors regain some order and tell us how to attach the ropes and the special action that will feed the rope through safely to allow our partner to climb and abseil down. Janet is worried by the height of the wall. The instructor directs her to the easiest climb and Emma encourages her by pointing out what might be a good route to try because the foot and hand holds are quite large and well-spaced. As Janet starts off there is a lot of support from the watching women – 'That's great!' 'You're doing really well' – and she climbs to right to the top despite her fear. High up the hesitancy returns as the instructor tells her to lean back. 'I can't do it...'. Janet begins but she regains her confidence when her friends promise that she can. There are cheers and congratulations as her feet touch the ground (field notes).

The women also built friendships from talking together at the coffee mornings and around the activities. Typical conversations involved sharing their everyday experiences of their new lifestyle and routines: shopping, appearances, cooking, visitors, trips home, children's homework, maids and husbands. In Chapter 8, I discuss how mum's groups in the city today have increased in number yet have many similarities in their aims.

Although being a member of these kinds of larger groups can bring a sense of collective belonging, there is also a desire for a different sort of interpersonal relationship. Janet and Emma, whom I accompanied to the climbing challenge, for example, had established a friendship that went far beyond the meetings of the Dubai Adventure Mums. As a result, an understanding of the function of the VCOs is not, I would argue, a substitute for understanding friendship as a significant intimate relationship in British migrants' lives. In the remainder of this chapter, therefore, I turn to explore how friendship is negotiated in transnational spaces.

Transnational Mobilities and Friendships

While the impact of global work on family lives is better understood, friendship has not been included in these discussions, echoing a broader devaluing of the significance of friendship noted by others, such as, in geography, Bunnell et al. (2012). The impacts of migration on friendship, at least those that became apparent during my fieldwork, were varied. In this section, I examine the 'challenges' described by my interlocutors: the difficulties of negotiating friendships, firstly, across transnational social fields and, secondly, among a highly mobile community of British nationals in Dubai. In the next section, I examine the more positive impacts of migration on friendships discussed by interviewees.

For some British migrants, the physical distances arising from migration from Britain to Dubai present challenges to the negotiation of their interpersonal relationships in the UK, including with friends. John, for example, had lived in Dubai for six years with his wife and they had had two young children since relocating. In an interview, he explained:

> You never get phone calls from them or letters from them [friends in the UK]. Keeping in touch is firmly placed as the responsibility of the people who have left. Here, it's hard to find real friends, deep-down, reliable friends, but there are about six sets of friends here that we will always keep in touch with, wherever we end up, because they understand what it's like to move. Sharing this experience with them too, of living here, means they have mostly replaced friendships from home. Especially since they also know our kids (interview).

John's narrative resonates with the sense that many Britons held that, because they are voluntary, career-driven migrants, their choice to migrate is frequently understood by those they leave behind to mean that they don't care enough about their personal relationships. However, John's narrative not only tells us about the challenges of maintaining friendships over distance, but also about the significance of time and the importance of sharing major life events, such as births, marriages and funerals, as well as 'being there' in a more everyday sense to share particular life stages. Narratives such as this from John remind us of the importance of time in the development and deepening of friendships, as well as how friendships can shift across the life course (Pahl 2000: 86).

In contrast, those British migrants who had moved frequently to new countries suggested their relationships with friends in the UK remained the most significant. An interview with James illustrates this:

> Hotel management tends to isolate you. You do a year here and a year there and you really can trace your good friends back to school and from that point you move around every couple of years and it's, you know, it's very hard to make solid friends. It is really hard to make real friends here because I know I'm not going to be here that long. You make some sort of shallow friendship, but nothing really what you rely on, you know (interview).

Lucy, too, was highly reflexive about the impact of her husband's global career trajectory on her own friendships. Rather than the challenges of distance, however, Lucy noted the impact of the sense of temporariness she felt in Dubai. This was frequently talked about in vague terms, but Lucy quantified it for illustration purposes, explaining that, in the six months before our interview, no fewer than sixteen families in her social circle had moved on to a new country or repatriated from Dubai. She also knew that her own residence in Dubai was based on the two-year employment contract of her husband, so she felt her family was 'in limbo'. It becomes evident, then, that the UAE *Kafala* sponsorship system shapes the intimate subjectivities of migrant workers and their families, even the friendships of more affluent middle-class 'expatriates'. Lucy explained:

We've just ended up scuba diving and our social life revolved around that and, of course, 'Adventure Mums'. So I've had a lot of things but all social, a lot of fun things. Just being a Jumeirah Jane for a couple of years. Coming here, I think I guarded myself a lot. I didn't want to be hurt again, so I think I was very careful. I didn't want to become too close to anybody because I knew it was for a short time. Really I was just protecting myself. I don't think I've changed my ideas of friendship, it's just, I suppose it shows you can have fun, you can do a lot of things without, but on a superficial level. I mean it is ok, for a while. The one thing I really miss here is that the friends I really confide in are all at home. Because I think people move on here, I don't feel so happy confiding in people. Personally, I very rarely confide in people here, but that's just me. I mean I've got probably two really good friends here, but, even them, I probably wouldn't confide in them as much as I do with my friends at home. I think you've just got the feeling that, even if you're really good friends with them, they might move on one day (interview).

For Lucy, the kinds of relationships experienced in larger social groups, as described earlier in this chapter, were distinguished from friendship, for they were based on 'a superficial level' of relation. Proximity with the 'expatriate community', then, has not led to a sense of intimacy among them. Rather, her temporariness, and theirs, has informed the way in which she orientates her intimate self through transnational social space, rather than towards friendships with other transnationals in Dubai.

Intriguingly, therefore, the perception of transience in Dubai impacted on what was deemed to be an 'appropriate' friendship. Stephanie, for example, suggested that a 'successful' friendship in 'expatriate communities' might be understood quite differently in terms of its character and enactment:

I like not having a complicated relationship – in the sense of being emotionally unbalanced, very clingy, she'd be in a huff if I did something without her – that's stifling. People who are always there for you, up for doing things with you, 'cos I can drift for weeks without bothering. *In these countries*, that's the best way to be: you need relaxed relationships because things can become very spontaneous. And you need to be flexible. It's almost etiquette to bring someone else along (interview).

Stephanie's narrative reveals the more generally recognised need in friendship that Rawlins refers to as a 'dialectic of freedom to be independent and the freedom to be dependent' (cited in Pahl 2000: 91), as she noted how this might shift in transnational spaces. Indeed, she indicated that the expectations, conventions and 'what is deemed to be an appropriate friendship' vary with spatial context (Bunnell et al. 2012: 494). These expectations and conventions, however, were often tacitly communicated, were not always shared, and were often experienced in emotionally unsatisfactory ways, so that Dubai emerged in the narratives of some British migrants as a city in which the habitual rules of friendship could not be taken for granted as they might be at 'home' in the UK. Claire's reflexive narrative of her experience of Dubai residence was both typical and evocative of this process:

> There's something you almost have to go through here: you think you have a really good friend and then people can switch off. If you talk to most people they'll say they had someone who they thought quite a lot of, and they thought was quite a good friend, and then all of a sudden they just don't want to talk to you anymore. And that happens a lot here, it's really weird. You know, I had someone I used to play tennis with every week and you know suddenly I'm not, I don't fit in with what she's doing, so you're literally just [pause] shoved aside. These other people won't even acknowledge your existence after a while. But you talk to anyone of my friends here and they will all say the same thing: that they had someone who suddenly decides they are superfluous to whatever you're doing. People here, I think, are quite superficial. A lot of people they come in and they have these ideas about status and, um, it's quite easy just to get used here. People I know who haven't been here as long as me, we're friendly because we've got a common interest. That's now I think, but that's probably because I'm quite settled and constant. I don't really need to go out and make friendships. I think when you first arrive you do tend to grab a bit: 'I don't know anyone, oh god, I need some friends...'. I think some people do that. You do see some people around and they just seem to be into everything, and then you find out they've only been here two months, and then you think: 'What are they doing?' And generally, this is a great generalisation, but generally these people are the ones who get people's backs up. Especially people who've been here a while because they think: 'What's she doing? She crops up everywhere.' But I can see why people would do it, that's how they get to know people. I think it depends on the sort of person you are. (interview)

Claire's story not only tells of her more specific experience of friendship (or lack of) with someone she played tennis with, but is also revealing of the broader efforts of making friends in a new place. Claire is able to tell this story, to me as a researcher, because she is now in the position of having close friends in Dubai; indeed she draws on their collective experience to convey to me the broader resonance of her illustration. As such, she speaks of the emotional and political dynamics of friendship: the difficulties of negotiating an interpersonal relationship when class status and length of residence cannot be simply set aside. Clearly the context in which friendships are enacted and experienced matters to their embodied and emotional negotiation. As Ray Pahl (2000: 44) concludes, having traced ideas of friendship from the philosophical writings of the Greeks and Romans through to its modern form, 'friendship has to be seen in historical and cultural contexts'.

As this section has demonstrated, residence in an 'expatriate' community stratified by class, gender and nationality, and shaped by the temporariness of a strict migration regime, may lead to British migrants rethinking the process of making friends and the meaning of friendship in their everyday lives. However, the reconfiguration of friendship in transnational spaces also allows for positive re-evaluations of the significance of friendship and such experiences are the focus of the next section of the chapter.

Friends as Family

In his text on friendship, Ray Pahl (2000: 142) suggests that friendship, 'perhaps more than any other aspect of our social lives, has eluded the attempts of social scientists to be classified and codified.' In part, this may have occurred because of the overlap between the practices of intimacy and emotion that constitute friendship and other kinds of relationship through which our intimate subjectivities are enacted. Consider, for example, how Fehr (2000) describes friendship as a close relationship which is 'an important source of meaning, happiness, enjoyment and love', a description that could also be applied to some marital or family relations. In contemporary British society, and beyond, the couple relationship has been elevated and prioritised (Jamieson 1998), leading to less attention being given to other intimate relations. Yet, for Pahl (2000: 62), friendships are not secondary in importance after marriage or family, as they sometimes appear in academic study, but are relationships worthy of study in their own right, with the potential to become 'deep, enduring and binding attachments'. In this section, I explore the moments when the intimacies of friendships between British migrants become described or experienced as 'family-like' relationships. What is it about everyday life in Dubai that shapes practices of friendship in such a way?

For married British women whose husbands travel frequently leading to regular absences, friendships can play a hugely significant role in the everyday routines of home, family and parenting. Kate's husband Trevor was a pilot, so he was frequently away overnight. She explained:

> A lot of friends here are very transient friends, ones that you'll only know in Dubai but you'll never know after, but then there are also some very, very, solid friendships that are built here. Ones that I could walk into a girlfriend's house and I know where everything is, I know their school patterns, I know her, what she'll be doing on a Monday afternoon with the kids, and that's not because we live on top of each other, ok we live five doors apart, but more because we rely on each other, we're emergency contacts, almost next of kin to each other, because we are in a country where families are not part of that, husbands either travel a lot or are not contactable, and therefore the wives become very…'community related'. We are our support…and there's a lot of things that we will discuss. And it sounds a bizarre way of life really, of family life, but we almost become a closer unit. When they are away you might get a phone call once in a day…I'm not going to waste it telling my husband that David [her son] has a cold. So all these sorts of things get talked about with friends (interview).

For Allan (1996: 108), contemporary friendships often share characteristics of familial relationships, particularly in the mutual favours through which friends may help each other in the daily routines of life; however, this can become exaggerated away from home where their supportive function may become even more apparent.

In the absence of kin support, the development of friendships may also be quicker than it might be at home. For example, Janet and Emma quickly formed a close friendship, and by extension linked their families, through early experiences of being able to help each other as new arrivals:

> Janet and Emma were in fits of giggles telling me how they got to know each other better. About ten days after they met, on leaving Janet and Graham's house at one in the morning, Emma found she was locked out and her own husband, Mark, was away on business. Janet's husband Graham drove to Emma's house and climbed over the six-foot garden wall, nearly landing in a six-foot hole the other side. Emma explained: 'I was thinking I don't know this woman and I've just killed her husband.' A week later, Emma and Mark were on their way to collect her mother from the airport at five in the morning when their car broke down. They had no breakdown service, so they hesitantly telephoned Graham again. Janet didn't see this as an imposition; explaining, 'We were just laughing at them, but the truth is things like that always happen to us normally' (field notes).

Without the wider familial support that they could take for granted in an emergency situation at home, the two women had formed a close bond. As Janet explained in a later interview:

> We've forged our friendship on disasters and embarrassment. Without relations, this is how we have to function. There's no one else there. [The narrative is then interrupted by Emma when she remembers that she needs to ask Janet whether she can look after her children later in the week and they speak between themselves for a few minutes to arrange the practicalities of collections before resuming.] It's a pressure and it does weigh on you, and you have to cash in favours from each other all the time (interview).

However, without their mutual liking, the demands on this friendship might have led to its demise. As Rawlins (1992) suggests, friendship involves a dialectic of affection and instrumentality.

Friendship is also hugely significant in the lives of single British migrants in Dubai. Conradson and Latham (2005b) observe single New Zealanders relocating to London with their friendship networks from home almost intact, such that friendship both sustains, inspires and shapes their mobility patterns. In contrast, single British migrants in Dubai, at least at the time of my fieldwork, rarely arrived to friends already living in the city. As has been noted of British migrants working in other global cities, collegial networks often provide initial contact points and after-work bar culture enables such relations to flourish (Beaverstock 2005; Scott 2007). However, the development of 'tribal-ephemeral' networks (Scott 2007) in pubs and bar spaces, as an extension to the workplace, is well understood in this wider literature and also evident from the discussion in Chapter 6, so here I focus instead on how shared households, and domestic spaces more generally, are key sites for the production of friendships. Indeed, the notion of friends as 'families of choice', whereby the distinction between family and friendship is blurred, is a productive one in this city (e.g. Bowlby 2011; Roseneil & Budgeon 2004). Homes are sites to which we can invite friends and, in doing so, deepen friendships through hospitality, care and time spent together. While this may also be relevant to married couples who may host dinners and parties, the importance of domestic space in socialising practices was highlighted in my analysis of the everyday life of young single professionals. Furthermore, in the case of shared households, homes can also be places where friendships can emerge, deepen, or end, in the negotiation of everyday domestic practice close up. During the time I lived in Dubai, many single Britons shared large apartments or villas, either choosing to sub-let a room or being placed in such accommodation by their employer. So, in this section, I consider the importance of shared households in enabling friendships, as well as the role of friendships in transforming our site of residence into home.

Paul, a journalist in his twenties, also lived in a shared villa and his friendship practices were illustrative of this wider phenomenon. I arrived to interview him about his sense of home and the domestic material culture he identified as significant was a DVD collection, including a box set of a British comedy series called *Spaced*. Paul explained that his friends from work had come for drinks at his villa before going out to a nightclub but, after initially deciding to watch a single episode, they forgot their plan and watched late into the evening. Watching television together with his friends had since become a regular way for Paul to spend Fridays, especially when he was hungover from a previous night's drinking. Similar occasions occurred in my own villa when trips back to the UK enabled housemates to purchase new series of *The Office*. Hanging out informally 'at home' to watch TV together was significant for single British migrants as they were behaving as a family might. The (illegal) purchase of films from Filipino and Chinese street-sellers was also routine but, arguably, there was something different about watching episodes of *British* television series such as *Spaced*, *The Office*, or *Ali G* and *Bo Selecta* on DVD in Dubai. Silverstone and Hirsch have argued that domestic technologies, such as televisions, are 'doubly articulated into our domesticity' through their simultaneous material and symbolic function: 'a means both for the integration of the household into the consumer culture of modern society – into a national as well as an international culture – and for the assertion of an individual's, a household's or an island's own identity...' (1992: 4). The setting and humour of these television programmes contributed to the informal shaping of intimate subjectivities among British friends, while other programmes, such as *Sex and the City* or *The Sopranos*, were more clearly consumed across the Western 'expatriate' community. The television officially available in the United Arab Emirates was limited at the time and even the South African satellite channels to which some households had access did not help reproduce a British sub-culture. As Morley (2000) suggests, today's sitting rooms are 'a place where, in a variety of mediated forms, the global meets the local', a place where people can experience displacement while staying still (Morley 2000: 11).

During the time I lived in Dubai, I was lucky enough to experience this informal socialising in shared villas and the friendships produced in household spaces. The small patio and garden of our villa opened onto a larger shared garden with a swimming pool and individuals or groups from other villas would often wander in from the garden, sharing food, cigarettes and alcohol spontaneously, or joining us in the living room to watch sports or a film. Friends of housemates would become acquaintances and then friends over time. I cannot overstate the significance of the simultaneously routine and spontaneous nature of these social interactions in the building of friendships: no invitations were necessary. We lived in each other's lives in a way that I have not experienced before or since. Sharing halls at university was somewhat different, not least because the living space was less homely and we shared for a matter of months. Some migrants may live for years in such a close community, with others moving in and out. Their domestic lives are therefore full of intimate friendships, contrasting with the challenges many experience in finding a romantic relationship in Dubai (as discussed in Chapter 6) and challenging some of the other accounts of friendship being made superficial by the temporariness of migrants' lives. The significance of British migrants' friendships thereby resonates, in spite of their relative privilege, with Elsheshtawy's (2010) observations of low-income migrant workers for whom gathering with friends is a vital part of everyday life.

Conclusion

With its connotations of fun, affection, caring and equality, friendship is a complex form of relationship, the significance of which has long been recognised by anthropologists and sociologists, but rarely attended to, until recently, in geographical scholarship (see Bunnell et al. 2012). Yet, the significance of friendship in Dubai was very difficult for me to ignore. On a personal level, this was a city in which I lived for only a short time, yet experienced some of the most emotionally significant and transformative friendships of my life. Meanwhile, ethnographically, my emerging research suggested that the conversations of those around me frequently focused on the negotiation of friendship in Dubai and the practices of the British migrants I was observing as they participated in social organisations and leisure spaces provided further evidence of the importance of friendship in their everyday lives. In this chapter, therefore, I furthered my exploration of intimate subjectivities by examining how transnational migration influences friendship practices, attentive to the city-specific narratives of the British participants in my research. I began by exploring the way in which the geographies of their friendships are shaped by the production of 'expatriate' subjectivities in Dubai, as identified in Chapter 4. Memberships of nationality-based VCOs, clubs, and privatised leisure environments enable and encourage Britons to form friendships primarily with other British nationals or, to a lesser extent, among a wider English-speaking community of affluent 'expatriates'. These spaces are important in making Britons feel at home in a new city, yet they also help to reproduce the classed and racialised stratification of Dubai.

The chapter continued by examining friendships more directly. I echo Conradson and Latham (2005b) in urging research on friendship specifically. If we focus only on clubs and institutions, then we miss out on a significant part of migrants' sociality that is, furthermore, not always captured by the related concepts of family and love. This chapter has demonstrated that friendships made in Dubai were vital to the intimate subjectivities of both married and single British migrants. Indeed, the strongly gendered division of labour in 'expatriate' households and the demands of mobility and long working hours placed on British men working in Dubai lead many accompanying women to look to friendships to support their parenting practices or to stand in for wider kin in situations of household crisis. For single British men and women, residence in shared households can also lead to 'family-like' relations that are produced in the sharing of intimate domestic space and routines. Like Eleanor Wilkinson (2014), then, I argue that popular focus on heteronormative coupledom in contemporary society should not obscure the significance of friendships as a form of intimacy beyond 'the family'. Chapter 6 further considers the intimate lives of single British migrants in Dubai, turning our focus towards sex, desire and romance.

Chapter Six
Sex, Desire and Romance in the Globalising City

After eating out with Malcolm and Ed, we go to a bar by the beach, then cross the city by taxi to go to Scarlett's in Emirates Towers. In the taxi, Ed, twenty-eight, comments: 'It's not a major city, so we make up for it with a lavish lifestyle.' He's employed in IT-recruitment, a job he calls 'corporate pimping, essentially' and confesses, with a boastful pride, to be 'shallow work, one of those jobs where you're making money out of fresh air'. One of Ed's friends rings him on his mobile and they arrange for our groups to meet in the next bar. Then we are interrupted by a text message he receives from a woman he met recently at the [horse] races, he explains: 'I bump into her every six months and we do it [have sex].' He laughs, but then he reminds me that he has a girlfriend of three years who lives in London, saying: 'I'm very much in love with her, I can't wait until we live together. I admit though, I'd hate her to play around and I'd end it if I found out she had. But, you know, it's hard living here, "out of sight, out of mind" and all that. I got laid [had sex] last week! I went out with a mate and we met up with some ladies and bought them champagne and pints of vodka tonic! It got very debauched, very Dubai!' He laughs and jokingly reflects: 'I'm such a player!' (field notes).

This chapter highlights couple relationships, focusing on the intimate subjectivities single straight British migrants enact and negotiate as they navigate Dubai's 'global nightscapes' (Farrer 2010), including ideas about the appropriateness of sex outside marriage, cross-national desire, and emotional attachments. Heteronormative couple relationships – those expressed in domestic cohabitation and marriage – are the focus of Chapter 7. Here I instead direct our attention to the perspectives of a sub-group of single British migrants with distinctive lifestyles and sexual subjectivities.

Transnational Geographies of the Heart: Intimate Subjectivities in a Globalising City, First Edition. Katie Walsh.
© 2018 John Wiley & Sons Ltd. Published 2018 by John Wiley & Sons Ltd.

In doing so, I engage with contemporary geographical perspectives on hetero-sexualities informed by queer theory (e.g. Hubbard 2008; Oswin 2008). Natalie Oswin (2008: 96) suggests that queer theorists might 'embrace the critique of identity to its fullest extent by abandoning the search for an inherently radical queer subject and turning attention to the advancement of a critical approach to the workings of sexual normativities and non-normativities'. From this agenda, there is space for critical attention to heterosexualities, too, not least their diverse enactments, power geometries and instabilities. This chapter demonstrates, for example, that heterosexualities are spatially contingent, with British migrants themselves noting that distinctive practices and experiences of couple relations have emerged in Dubai specifically, and 'away from home' more generally. Attempting to move forward our understanding of sexuality in migration con-texts, Manalansan (2006: 2) adopts a queer theoretical perspective, putting for-ward 'the tools of queer studies as a way to complicate and reexamine assumptions and concepts that unwittingly reify normative notions of gender and sexuality'. To illustrate this radical potential, Manalansan (2006: 11) offers a 'queer reading' of a sub-literature in migration studies on Filipino domestic workers, providing critical insight into the way in which heteronormativity impacts upon hetero-sexual migrants' lives (as well as how it has implicitly structured scholarly analysis of migrant subjectivities and practices) by unsettling normative conceptions of parenthood, maternal love, and care and repositioning migrant domestic workers as 'viable desiring subjects'. Importantly, in Manalansan's work it is the theoretical approach, rather than the subjects themselves, that is conceived of as 'radical': given the intersectionality of migrant sexualities with other social identities (e.g. race, class, gender), the non-heteronormative cannot be automatically celebrated as progressive (see also Oswin 2008). This is especially pertinent in this chapter, where British migrants' intimate subjectivities are shaped by the privilege of their whiteness and affluence. As the field notes of my night out with Malcolm and Ed above suggest, this is especially the case in the production of migrant masculini-ties in Dubai.

In this chapter, then, I consider how heterosexualities are reconfigured by migration (see also Farrer 2010; Huang & Yeoh 2008; Shen 2008; Walsh, Shen & Willis 2008). I begin by exploring the impact of the UAE's decency laws that pro-hibit sex outside marriage. I suggest that these are widely ignored by single Britons but that, nonetheless, they shape reproductive rights of British women in ways that remind us that migration is lived through the emotional body (see also Dunn 2009). I then consider how the racialisation of Dubai's urban 'landscape of desire' (Bell & Valentine 1995) both limits and informs instances of intimacy across boundaries of 'race'. Subsequently, I explore further the impact of the 'temporariness' of Britons in Dubai since, as argued in earlier chapters, British migration to Dubai is understood as a relatively impermanent relocation due to the *Kafala* sponsorship system, rather than as emigration, and this in turn informs the way in which my interlocutors approached relationships established in the city. In terms of couple relations specifically, I argue that geographies of displacement create opportunities for straight migrants to enact new intimate subjectivities that contest heteronormativity, echoing a socio-spatial dynamic that has been observed in queer migrations (e.g. Gorman-Murray 2007; Knopp 2004). Finally, I turn to consider the related question of how British migrants talk about romance and attachment in this global city. Throughout the chapter, I highlight how these intimate subjectivities are not only about performing new sexual identities, but are also gendered. The chapter, therefore, builds on those preceding it. It further explicates the production of expatriate subjectivities, introduced in Chapter 4, through attention to the enactment of nightlife geographies and inhabitation of privatised bar and club spaces, and I note the formal and informal attempts to regulate cross-cultural intimacies and their transgression. Further, the significance of the friendships among younger single Britons, as discussed in Chapter 5, becomes more apparent as the chapter proceeds, in particular the way in which friendships are experienced and enacted in relation to gendered, classed and racialised performances of heterosexuality.

Heteronormativity and the UAE's 'Decency Laws'

All sex outside marriage is considered illegal in the UAE, but individuals would not be prosecuted unless it were to come to the attention of the authorities, something that usually results only from three circumstances. Firstly, in cases of complaint, for example, employers' complaints against their domestic workers' relationships or cases of adultery. Secondly, in cases of rape, with the victim of the crime being prosecuted and abortion also strictly prohibited. And, finally, in cases of pregnancy, since expectant women are asked to show their marriage certificate in the course of accessing maternity treatment in hospital. Pardis Mahdavi (2016) explores the impact that this legislation prohibiting sex outside marriage – *zina* – has on low-income migrant workers in the GCC countries, collecting harrowing stories of imprisonment, deportation and statelessness. The British migrants I met between 2002 and 2004 complained about the lack of clarity over such laws, but there is now, at the time of writing, a very clear statement on the Dubai government website:

> It is of the utmost importance for a woman to be married if pregnant in the UAE. At the hospital when you go for your first check-up, you will need to show an original marriage certificate along with copies of your passport and visa. If you are unmarried and pregnant, you should either get married or expatriate unmarried expectant mothers should return to their home countries for the delivery (Government of Dubai 2013, cited in Mahdavi 2016: 52).

Furthermore, the UAEs decency laws prohibit sexualised contact with the opposite sex in a public place. The values of heteronormativity – marriage, commitment and monogamy – are therefore encoded into the city's legislation.

In contradiction to the severity of the declared penalties, however, I met unmarried British couples, gay and straight, who felt safe living together in Dubai. Furthermore, as later sections of this chapter demonstrate, sex outside marriage was widespread beyond cohabiting couples and rarely deemed a risk. Aside from some high-profile cases of arrest since my fieldwork ended, single Britons living in Dubai have not been criminalised for their behaviour. Perhaps this is because bars and clubs are located within hotels where private security operates, and the distance between hotels displaces public displays of drunkenness from the streets into taxis. Also, the UAE does not have religious police like its neighbour Saudi Arabia and, although the regular police have a Moral Guidance Department, the enforcement of decency laws is not widespread unless the behaviour in question is of offence to an Emirati national. So, British migrants were rarely inhibited by these regulatory controls, but they did express concern as to the ambiguity of the legislation. In 2009, after my sustained period of fieldwork had ended, one Briton was prosecuted in a legal case publicised by the international media: Michelle Palmer received a three-month jail sentence when found guilty of having sex on the beach with visiting Briton Vince Acors whom she met at a brunch. Complaints about the perceived ambiguity of the decency laws then led to a publicity campaign in which a series of articles were published in the English-language newspapers detailing what would be considered appropriate and sensitive behaviour.

There were, however, moments during my fieldwork when the decency laws and their ambiguities had an observable impact on the experiences of British migrant women. The intersection of race, class and migrant status with gendered migrant subjectivities requires closer examination in relation to their embodied and affectual experience. Just as the decency laws concerning dress impact more on British women's bodies than men's so, too, do those focused on reproductive rights. This is demonstrated with this example from a narrative about the fear of an unwanted pregnancy from a British woman in her twenties:

I looked through the *Explorer* (local directory) for doctor's surgeries and chose the one nearest to me. Ringing for an appointment, I was shocked by the response: the receptionist said 'no, sorry, it's illegal here, but come in anyway, there are ways we can help you'. It all felt so sinister, it made me think of knitting needles, so I was feeling a bit panicky by now. So I phoned the clinic with British-sounding doctors' names and spoke to an English-sounding receptionist. I blurted out: 'I need to take 72-hour contraception but they told me it's illegal.' She was calm and said, 'yes it is, but we might have some', so I booked in. By the time I went to my appointment a few hours later, I was in quite a state because she'd been so non-committal on the phone. All these scenarios were flitting through my head in the waiting room. If I couldn't get it, would I have to fly home to have an abortion? My friend had to do that, but first she had to wait for holiday leave and her husband's relatives to visit, so it was awful. After seeing the nurse, a doctor gave me a prescription for the normal contraceptive pill and told me to take seven in one day. I asked for anti-nausea pills, but she said she couldn't give them to me [as] it would be too obvious. She gave me travel-sickness pills instead. They sent me to the pharmacy next to [*text removed*] as they're used to it from the flight crew [and so] they'd 'turn a blind eye'. But the Indian pharmacist looked at me like I was a whore! (field notes).

As this example suggests, the unmarried British woman's body is subject to the attempted regulation of sexual intimacy not only by Dubai's decency laws, but also by the surveillance of the medical profession. This legislation is clearly gendered since, as Mahdavi (2016: 16) puts it: 'pregnancy makes women's sexuality immediately visible'. This is a rare moment where the normative status of whiteness as privileged is destabilised and the British migrant woman shares something of the fear that many low-income migrant workers from Asia encounter. At the same time, however, and as the statement from Dubai's government makes clear, pregnancy outside marriage can, for the affluent 'expatriate' with resources, be solved by accessing emergency contraception, marriage, or by returning to their home country for an abortion or the birth.

For single straight British migrants in Dubai, heteronormative values are further espoused by the British expatriate families they find themselves living alongside. The difference in lifestyle between the single and married Britons in this context should not be underestimated. As described in Chapter 4, in some cases the urban mobilities of these two groups may rarely overlap beyond the workplace and they may inhabit almost entirely separate social spaces and residential neighbourhoods. This distinction is vividly evident when comparing married and single women's lives. In their work on Singapore, Yeoh and Huang (2010) have identified that migrant subjectivities are not only strongly gendered, but marital status emerges as significant in how women's identities are understood. They explain how the 'trailing spouse' is perceived as having a place in the city by virtue of their status as 'wife' to expatriate men, whereas 'unaccompanied migrant women are seen to be "out of place", and therefore either vulnerable and easily victimised, or predatory and potentially dangerous' (ibid.: 39). The single women they describe are study mothers and domestic workers who might well be married but, since they must leave their husbands behind, inspire certain anxieties within local society because of their perceived potential to destabilise heteronormativity. Likewise, it is single British migrant women whose presence in Dubai is perceived as more visible and whose bodies become the site of negotiating the values of heteronormativity. I explore this further in the next sections, while leaving discussion of heteronormative British family life to Chapter 7.

Race, Affluence and Masculinity in Transnational Space

The social hierarchies structured by citizenship, race and class, as described at length in Chapter 3, as well as the segregation of urban space further explored in Chapter 4, also help to shape performances of migrant sexualities that emerge in

Dubai. In this section, then, I focus on how British migrants' sexual subjectivities, their heterosexualities in the case of my research participants, are reconfigured by the meaning of whiteness and its intersection with migrant status and gender. It is worth noting that the status of the Emirati nationals makes them unapproachable as potential short- or long-term partners for migrants, even for well-connected British residents. There were a small number of middle-aged British women who had married Emirati men they met in the UK but these were an exception: fears about cross-cultural intermarriage are thought to be part of the rationale for the *Kafala* system across the GCC countries (e.g. Kapiszewski 2006) and, as discussed in Chapters 3 and 4, wealth, religion and citizenship status mark these communities as distinct. At the same time, conscious of their relative privilege in both a global and local social hierarchy based on nationality, race and class, and seeking segregated leisure spaces and institutional lives (see Chapters 4 and 5), British men and women tend not to seek sex or romance with migrants of other nationalities beyond the larger 'Western' migrant community. However, there are some exceptions – notably, British men's relations with sex workers and domestic workers – and in this section I examine how these are framed by the men themselves and the wider British community. I argue that these are relations that draw on racialised notions of masculinity and which demonstrate the significance of heterosexualities in the remaking of British migrant subjectivities in Dubai. Finally, I explore the way in which the positionality of British women is more ambiguous, including in those moments when their status is unclear and they may be confused for sex workers.

Arriving in Bur Dubai to stay at a hotel during my first visit to Dubai, I became immediately aware of the presence of the sex work industry within the city. In this central district, it was not uncommon to see sex workers waiting on the streets and several hotel bars, such as British-styled pub 'The York', were populated by sex workers and their male clients. Pardis Mahdavi's (2011) sustained analysis of the sex worker industry in Dubai supports my initial observations of not only the overt visibility of this industry within the city, but also the ethnic diversity of the clients. She notes that Emiratis are under-represented among the tourists, business visitors and expatriates of all nationalities who dominate these bars. In comparison with other topics, I collected little material on British men's relationships with sex workers, since they were rarely spoken about in front of me and other women. This is partly because it is seen as morally unacceptable for men to engage with sex workers for fears of such women being trafficked and/or exploited, but also because it raises the possibility that the consumer is unattractive to other co-nationals and wealthier migrants. Exceptions to this include boasting about employing several sex workers simultaneously and other forms of 'fantasy' sex. The discursive framing of Dubai as a 'holiday-like' space that I highlight later in this chapter extends to how some British migrant men were able to explain their interactions with migrant sex workers in ways that didn't diminish their masculinity. One episode in my fieldwork politicised the sex work economy in the daily conversation of my household and I turn to explore this now as an example.

The 'laddish' culture of my shared house (at that point I shared with four young men) had previously been limited to an FHM calendar behind the bar and occasional sexist banter (not directed at me, but to some extent unhidden from me). The organisation of a 'gentlemen's evening', something with which one man in the house and myself were not involved, was much more controversial. Planned in jest over many drunken nights, when it actually happened the invitation read:

Gentlemen,

> Bar Smith requires the pleasure of your company for an evening of gentlemen's pursuits – namely, the consumption of alcohol and gambling of money in the company of scantily clad young ladies.
>
> 2 card tables will be running, operating 3 Card Brag and Pontoon to the Bar Smith house rules (a brief synopsis is attached).
>
> There will be a third table, operating dealer's choice – the game can change or remain the same as the deal goes around the table.
>
> As it will be St George's Day, any player leaving one table to join another is required to toast Her Majesty the Queen.
>
> Players are required to bring the following:
>
> - AED 50 – contribution to 2 professional ladies to serve drinks from 9.30–1.30 am*
> - AED 150 – minimum playing float
> - An alcoholic contribution of choice (i.e. a bottle of spirits or a slab of beers)
> - collar and tie dress code
> - cigars (optional)
>
> *A note on the ladies: they are hired to serve drinks till 1.30, anyone interested in 'extras' is welcome to do their own bartering once they've poured the last vodka tonic – but my door will be locked!

I spoke to my housemate Ed before the event and he explained:

> Everyone chips in fifty dees for the hire of two prostitutes. I've made it very clear to everyone the prostitutes will be doing nothing but serving drinks, clearing ashtrays, smiling sweetly and maybe topless, but I don't know, that could be a little seedy. Basically I'll sell it to them that this is a great opportunity to get three hundred Dees (UAE Dirhams) for 4 hours' work and they don't even have to have sex, just fawn over us and serve us drinks. Then they will be free to go on their usual haunts, but there is always the possibility they could pick up extra money here. Basically it's a boy's jolly to pretend that we're a bunch of high rollers and players, when we're really just monkeys essentially, and because we can in Dubai. It's just a bit of a laugh. And the toasting the Queen thing, it's just a bit tongue in cheek (interview).

The field notes, as well as Ed's interview, reveal how the discursive framing of British migration to Dubai through 'expatriate' subjectivities can lend itself to the production of a 'Western' hypermasculinity (see also Farrer 2010). In the episode described above, the men's fantasies clearly drew upon long-established Orientalist and colonial tropes of the subservient female 'Other' as being sexually available for the British man abroad to fulfil his desire (Hyam 1990: 9; Said 2001). This kind of masculinist travel fantasy has been strongly critiqued in the context of sex tourism (e.g. Seabrook 1996), yet Ed repeatedly downplayed the unequal and racialised power involved. Despite revolving around fantasy, this vignette is also about ordinariness: the event took place in a domestic setting, it involved very typical young British men – not sexually desperate, not physically unattractive, and not particularly rich – and it was an 'opportunity' they understood as being much more easily available to them in Dubai than in Britain. As Sanchez Taylor argues in relation to sex tourism:

> As far as white male sex tourists are concerned, it is not just cheap sex they pursue. They also like travelling to 'Third World' countries, where they feel that somehow the proper order between the genders and 'races' has been restored […] sex tourists find that their masculinity and racialized power is affirmed in ways that it is not at home (Sanchez Taylor 2000: 43).

For the men involved in this vignette, their performance of a particular type of white masculinity is constructed not necessarily through sexual relations themselves (many of the men only participated in the card games), but through the empowerment of knowing they could engage in paid relations if they so desired. Since young white British or expatriate men rarely expressed a desire to enter a relationship with migrant women from Asia in Dubai, British women did not experience the sexual disadvantage or anxieties of dating they describe in Asia (see discussion of Filipino migrant women later in this section, as well as Farrer 2011; Lloyd 2017; Willis & Yeoh 2007).

It is this sense of the racialisation of Dubai's cultural politics enabling a hypermasculinity to be performed by white British men that also shapes another set of intimate sexual relationships in the city: those of married Britons. While I examine their constructions of family and home in the next chapter, here I wish to comment briefly on the phenomenon Shen (2008) has termed 'situational singles', whereby transnational migrants living away from their wives perform a hypermasculinity in response to their increased financial resources and social status. During the early 2000s it was common for British women, especially those with children on summer vacation from school, to return to the UK and stay in their own homes or with relatives. In part, this was to escape the oppressive dry heat of summer in Dubai where temperatures can reach 50 degrees centigrade, but it also served, as 'home leave' would have done for colonial expatriates in the past, to connect them with their British homeland. This 'summer exodus', Frances explained one afternoon, 'leaves all the men rattling around in these big villas by themselves' since they remained in Dubai to work. Frances continued to explain not only the impact upon marriages between Britons, but also how the men responded: 'It's not good for a marriage to have three months separate, is it? It's asking for trouble. And, a lot of the men just look elsewhere. It's handed to them on a plate, of course' (field notes). The masculine subjectivities of these 'summer bachelors', as I heard them referred to, resonate with the 'situational singles' Shen (2008) identifies among Taiwanese business men working in Chinese economic zones who, away from their wives and families in Taiwan, enjoy the attention of Chinese sex workers and girlfriends resulting from their increased economic status. While no British woman confided in me that their own husband had had an affair, I was frequently told of an urban legend by interlocutors: 'the year that a whole group of British women returned with their children from an extended summer holiday in the UK to find their husbands had replaced them with Filipinas [sic]'. Migrant women from the Philippines, predominantly working in lower-skilled positions in the retail, hospitality, tourism and domestic work sectors, tended to be considered socially inferior by British women. Yet, the anxieties that British women revealed in their stereotypes of Filipino migrant women being 'after a husband' centre not only on the relative youth of the Filipino migrants, but also on wider anxieties about body size and femininity that expatriate women experience in Asia (see also Farrer 2011; Lloyd 2017; Willis & Yeoh 2007). The perception of younger, single, 'foreign' women being in 'competition' for Western men and a corresponding sense of vulnerability and insecurity is perhaps partly responsible (alongside a higher disposable income, cheaper services, more leisure time) for the heightened emphasis on grooming and fitness among married British migrant women in Dubai.

At the same time, British women appear completely uninterested in establishing relationships with non-expatriate male migrants, especially those from south or east Asia. While Farrer (2011) identifies a 'racial barrier' on both sides preventing sexual relationships being established between white Western expatriate women and Asian men in Shanghai, my observations suggest that in Dubai the racial distancing was primarily the result of British women's choices to decline the attention of non-Western men. For example, my field notes recall occasions when the nightclub Zinc was dismissed as the next destination because its clientele was considered too ethnically diverse and, as such, 'there's not going to be anyone we want to sleep [have sex] with' (Felicity, field notes). Instead, the Lodge was chosen as somewhere more likely to provide opportunities to meet with other Britons and the wider Western expatriate community. Although nightclubs are, indisputably, an 'interethnic contact zone' (Farrer 2011), they were also highly regulated spaces where contact across nationality was discouraged. For example, where men refused to observe the tacit rules of segregation that implicitly structured dance-floor intimacies, they were ignored or ridiculed.

British women in Dubai described the Asian and Arab middle-class men they encountered as 'letchy', a term which is suggestive of an insistent gaze and unwanted tactility that sexualises the recipient. It also indicates perhaps a transfer of 'bachelor fear' (see Smith 2010) from the bodies of low-income unaccompanied migrant workers described as 'bachelors' (even though they may have wives and children at home) onto middle-class migrant men from Arab and Asian countries. British women based these racialised stereotypes on their frequent experiences of being propositioned in Dubai when not traveling by car or when wearing swimwear on a public beach. This extract from my field notes records the experience of Felicity (teacher, late twenties) on her way to visit my own place of residence in Satwa one evening:

> I get taxis everywhere right, and then tonight I thought, I'll just walk: it's only five minutes round the corner. First, I had this Indian man video-taping me as I went past and I couldn't hide anywhere or get him to stop. And then just as I was turning into your road, this Arab in a Land Cruiser drew up and fucking asked me 'how much?' You know, dish-dash [British slang for Emirati national dress] and everything, they can get away with it. Tosser! I just ignored him and didn't even look at him but he drove along slowly, so I told him to 'piss off!' He just smiled and drove off (field notes).

Unlike in Saudi Arabia, women in Dubai are legally free to dress as they please and, more recently, a telephone hotline has been established that women can call if they are stalked or harassed (Smith 2010). However, it is evident that if British women dress as they might in the UK, at least in certain districts, the status of their Britishness that protects them from working as sex workers in Dubai is not recognised by those they encounter. As the next section suggests, the vast majority of dating and sexual relationships enacted by young single British migrants take place within the British, or at least white expatriate, community in Dubai.

Dubai's 'Global Nightscapes'

For James Farrer (2011), observing the production of sexualised identities for skilled migrants in Shanghai, the 'global nightscapes' of the contemporary global city figure as a site of particular significance for co-presence and sexualised encounter between migrants. As in Shanghai, the nightscapes in Dubai consist of both 'enclaves', where Britons can seek comfortable affiliation among other 'expatriates', as well as 'contact zones', in which they might encounter a range of differently racialised bodies (Farrer 2011). Bars and clubs are, therefore, sites of consumption that are simultaneously local urban spaces and sites of transnational flows (Farrer 2011: 749). Furthermore, in Dubai, such spaces provide relatively unregulated sites of liminality, where the more modest dress codes British migrants follow in spaces where they mix with Emirati nationals can be ignored. In this respect, for nearly two decades now, Dubai's nightlife has been noted in English-language guidebooks and the British media as exceptional within the GCC region. In an early edition of *Culture Shock! United Arab Emirates: A Guide to Customs and Etiquette*, for example, Dubai was described as 'the Gulf playground' (Crocetti 2000: 49). Similarly, an article published in both *The Daily Mail* (a tabloid newspaper) and *Cosmopolitan* (a magazine directed towards young women) called Dubai a place of 'ultimate escapism', where expatriates are 'seduced by the party lifestyle' (Brinkworth 2001a: 20; 2001b: 56). The UAE has legislation prohibiting alcohol consumption outside licensed premises, as well as criminalising drunkenness in public, but alcohol is obtainable by those who have a licence to drink at home and is widely served in pubs, bars, restaurants and nightclubs. The weekly UAE-based English-language magazine *Time Out Dubai* and the website of the same name listed multiple pages of nightlife venues during the early 2000s.

In examining the way in which these spaces function in the production of migrants' intimate subjectivities, it is important to recognise that drinking spaces vary (Jayne, Holloway & Valentine 2006). My observations took place mainly in ten bars/clubs which were highly frequented by single British migrants at this time – Aussie Legends, Boston Bar, Boudoir, Jimmy Dix, Long's, Rock Bottom, Scarlett's, The Irish Village, The Lodge, and Zinc – and which played UK chart music, rather than being part of a more serious 'music scene'. Even though alcohol consumption was important in these spaces, they were not 'pubs' (e.g. The Red Lion or The Old Vic) which were instead associated with Dubai's older and/or married Britons. With the exception of Boudoir and The Lodge, at this time there were few affluent south Asian migrants using these spaces, in contrast to more mixed bars/clubs. These bars and clubs had racialised door policies and, therefore, were not as highly frequented by women sex workers or, more importantly, their clients (see also Mahdavi 2011). Some other establishments were known to be sites for men to 'pick up' sex workers and therefore avoided by British women, not least because they might be mistaken for a sex worker themselves. One manager explained that he was 'unable to discriminate, but in an ideal world [name of bar/club] would be pitched at your expat market 100%' (interview). Another explained that all clubs/bars have a 'member's only' policy and signage that could be invoked if necessary: 'We don't let in the cheaper ladies: we have a very strict policy on hookers…because of the quality of this product and because [name of bar/club] is such a strong brand, we just can't have people walking into our bar and thinking there are loads of hookers here' (interview). These bars/ clubs *are* highly sexualised spaces – Jimmy Dix, for example, was described in 2006 as an 'extremely popular meat market' by *Time Out Dubai* – but the privatisation of these spaces allows them to be highly managed and stratified by class and nationality, even while they are also 'contact zones' (Farrer 2011). These bars and clubs were also extremely competitive with each other for the 'expatriate' market, leading to frequent offers and promotions, such as all-day 'brunches' where you can eat and drink as much as you like, and 'ladies nights' where women ('expatriate' women) were provided with vouchers for free drinks on the door. Such policies, along with the higher disposable incomes, led to the 'global nightscapes' (Farrer 2011) of Dubai being highly accessible to British migrants, shaping their intimate subjectivities 'away from home'.

Indeed, single British migrants went out more frequently and claimed to drink more excessively in Dubai than they had previously in the UK. The following field notes are a vignette of a typical night out:

> Encouraged by text messages I receive throughout the day – 'still on for ladies night? 10 ok 4 u?' – I take a taxi to Boudoir. There's a long queue, but I'm ushered to the front by the bouncers. Inside it's still quite quiet and I spot Tanya easily by the bar, but Tuesday's are 'ladies night' – a promotion that offers free champagne and straw-berries for women – and it's soon packed. The DJ is playing R and B: 50 Cent's *In Da Club* comes on and then they play Blu Cantrell – 'you say you love me, say you love me, but you're never there for me…if love hurts, it won't work, maybe we need to some time alone, we need some time to breathe'. The crowd goes mad re-living Cantrell's live appearance in the club a month previously, when she was standing on a table singing the track and pouring champagne into people's glasses after relocat-ing from Media City (where her planned concert had been cancelled due to non-payment). I bump into a trolleyed [very drunk] Vicky in the ladies' washroom. She tells me it's her friend's birthday and they've been to Spice Island, a restaurant with a permanent 'all you can eat and drink' offer: 'I didn't eat hardly nothing!' she reports. 'The waiters were getting really pissed [annoyed] with us because we kept ordering rounds of tequilas all the time.' I rejoin my friends and we also order shots, quickly downing the small glasses of sambuca or tequila, before continuing more slowly with our champagne. We're quite drunk now and the conversation moves on to men. Jane is thirty-one and a hairdresser. Originally from Chester in England, she has lived and worked in Spain and Australia. She hands round a packet of Marlboro Lights [cigarettes] and suddenly laughs, saying: 'Me pants have been round me ankles more times in Dubai than in the rest of my life put together!' (field notes).

The discursive framing of Dubai as a 'holiday-like' space shapes cultures of inti-macy in respect to the way in which single migrants approach their sexual/romantic encounters and relationships. Both an increased disposable income and the UAE's climate mark British migration to Dubai as partly about accessing a lifestyle that is distinct from that available to them when they were working and resident in the UK. Thereby, the geographical location of Dubai 'away from home' is a key reference point shaping the cultures of intimacy single Britons encounter and help reproduce. A belief in the sexually liberating or thrilling aspects of travel is deeply encoded in the discourses of desire and distance (Kaplan 1996). As Craik (1997: 127) suggests, the tourist desires to 'escape the usual disciplines and norms of everyday life by indulging in hedonism, fantasy and unrestrained sexual encounters during the course of a holiday'. I was struck by the repeated accounts of a sense of 'unreality' that surrounded British migrants' residence in Dubai, irrespective of their length of actual stay. Everyday life was described as 'like living in Disneyland', a holiday analogy evocative of excitement, leisure, escape, freedom and play. British migrants linked this analogy with the reconfiguration of the practices of sexuality and romance they observed or engaged in: a shift from the dating and monogamy seen as acceptable and appropriate among the middle class in the UK to frequent, episodic, sexual encounters they associated either with the working classes in the UK or the exceptionalities of holiday spaces.

Geographical imaginaries of movement and displacement seem, therefore, to strongly inform and shape cultures of intimacy in Dubai, both for British men like Ed, with whom we started this chapter, and for women like Jane (previous extract). The reconfiguration of cultures of intimacy that make it more acceptable to engage in frequent sexual encounters against the norms of white middle-class heteronormative discourses was clearly celebrated by many of the single British migrants I observed, irrespective of gender. This was also the case when they claimed to have enacted a much more conventionally settled performance of heterosexuality in the UK, and in these cases migrants framed their practices in terms of 'escaping domesticity'. Migration afforded an opportunity to escape from the 'domestication' of either their own relationship or that of their friends, as cohabitation became normalised among their peers. In doing so, they were able to narrate a construction of 'self' based on independence, spontaneity and openness, as can be seen in Jane's interview narrative. Jane was single and came to Dubai to work as a stylist for an international company of hairdressing salons. Yet, her narrative of relocation is not about her career, but framed instead in terms of intimacy:

> I got bored. I got itchy feet. A job come up and seeing as I had no commitments or anything. I've got to get it out of my system. Dad says I'm tied to a shooting star! I'd got to the point where it was either buy a house, make a decision about Gary [her boyfriend at the time] and settle down, Or, I do this [move to Dubai] and see if it's what I want to do. There's nothing holding me back. I always say that I don't think I'll settle in the UK. At the moment I'm just, like, I'll take it as it comes and see what happens. Travelling has made me a lot more independent. I have a lot more confidence. Now it wouldn't bother me going anywhere on my own. Yeah, I love travelling. It opens your eyes to different cultures and that (interview).

Jane was not unusual in this. Nor was this narrative of 'escape' unique to women. Craig, also in his early thirties, explained his own migration in similar terms:

> I moved here because I was ambitious, but also because I wanted to escape my woman troubles. Basically, I was involved with a woman, more like a fiancé I suppose, as we were ready to get married, and then I thought: 'No fucking way! It'll kill me!' My intention to be away two years or so has rapidly turned into eight now. I wanted total space; I didn't want to think about her. As soon as I left, I felt comfortable. It's almost like an on-off switch. Geographical: I was here, she was there. Yeah, absence makes the heart grow fonder, but distance makes it weaker. Even with wonderful communication devices, you can't do anything, so you get over it a lot quicker (interview).

Felicity also associated mobility with escape from the everyday mundanity of her previous relationship:

> I had a boyfriend until I moved to Dubai two years ago. We met at university when I was nineteen and were together eight years. When I look back to how I spent my twenties! We'd got into a real routine of doing things, like the weekly shop after work on a Friday, and I never went out in the evening after work. I was in a rut. He got a job in Dubai and I felt I had no choice but to follow him. After a month or so, his job fell through and he had to move back to the UK, but I decided to stay. After he left, I got a whole new lease of life. I consider the move a lucky escape. My friends from home who've visited me here say I've really changed: I'm back to my nineteen-year-old self. I went through quite a few men immediately afterwards, just having fun (interview).

Like extra-marital affairs (Kipnis 2003), 'promiscuity' is positioned in these narratives as a challenge to the widespread privileging of marriage and/or long-term monogamous relationships. These migrant heterosexualities resonate with other instances where casual sex is enabled by migrants being away from home. Malam (2008), for example, suggests that Thai migrants relocating to work in the beach bars of southern Thailand feel able to engage in casual sexual relationships with European backpackers because they are free from the surveillance and governance of their village communities, while Gaetano (2008) suggests that for rural Chinese women migrating to Beijing, their new sense of agency enables them to challenge parental authority and enter freely chosen love marriages.

In my field notes I also recorded both men and women describing their feelings of alienation in the UK when their friends began to enact performances of 'domesticated' heterosexuality they could not identify with:

> Today I listened to Ben, twenty-seven, talking about the relationships of his friends back home. He told me about his best friend who had been 'seeing' a woman for the last two years. Ben doesn't like her and can't believe how she stops his friend from doing things with Ben and their group of mates, like going away for the weekend. He moaned that she often expects him to stay in or 'gate-crashes' what Ben sees as boys' nights. He got quite worked up and described his friend as being wrapped around his girlfriend's little finger. He's worried that they will marry and their relationship will become like that of another married friend with children who now feels trapped (field notes).

> I was having a cup of tea with Julie in the garden after work. She left London when she was twenty-eight. She described how all her friends were settling down, buying houses, planning weddings, and starting to have children, and she just couldn't bear to stay among them. She told me how all they did was talk about their latest DIY projects in 'far too much detail' and that, even now [two years later], she dreads getting emails full of photos of a new baby 'looking the same as any other baby' (field notes).

For many of the single Britons I interviewed in Dubai, travel offered a welcome opportunity to make new friends who were also single and embrace an alternative lifestyle to that of their peers in the UK, many of whom were settling down. Mobility then, offers the potential to recast intimate subjectivities for those who are single. This is understood as being necessary because being single, especially being a single woman, still attracts much attention from 'concerned' relatives and coupled-up friends in the UK (see also Wilkinson 2014). Members of the Dubai-based Bridget Jones club, for example, told me that they most identify with Bridget Jones when she attends a dinner party of 'smug marrieds' and is forced to explain her single status (field notes). Furthermore, single British women in Dubai usually call themselves and each other 'girls' in conversation, an unconscious label that perhaps highlights their single status (without regard to age) and evocatively contrasts their own femininity with that of British expatriate wives and mothers. Like the 'player' masculinity, the 'expat girl' is becoming a paradigmatic figure of 'Western' transnational sexuality through which the city is 'made legible' (Hubbard 2002). So, the motivation to migrate can also be understood in relation to the persistence of powerful discourses in the UK about what type of lifestyle one should be leading at a particular age.

Dubai: No Place for Romance?

Co-existing with the celebratory enactments of displacement described in the previous section were those discourses circulating among British migrants in which these same performances of heterosexuality were identified as problematic, even by those involved. Many distinguished between their own practices and those of relative 'newcomers'. Recent British migrants were often described with terms such as 'fresh off the boat' or 'freshers', which is the term given to first-year students upon arrival at university in the UK, when they are expected to participate in a week of drinking excessively when away from home for the first time. Britons tended to distance themselves from the 'promiscuity' of the initial period after their own relocation:

> Tanya is thirty and from Sheffield. She has lived in Dubai for just over a year, working as a journalist. She told me that she went a bit mad when she first got here: going out all the time, getting really pissed [drunk]. She said she had sex with too many men in a short space of time, not that many, but far more than she would normally. I didn't ask her how many was too many: she was too embarrassed about it. She claims that she is now making an effort to date men without having sex, 'to get to know them first' (field notes).

Among those who were in an established couple relationship, whether it began in the UK or Dubai, the sexuality/dating culture was described as 'predictable', 'boring' or 'desperate'. Such concerns resonate with those of some social theorists who fear the wider depersonalisation and anonymity of the modern world in their analyses of love (e.g. Bauman 2003).

In Dubai, such fears are articulated through ideas about the transnational character of the city and its transient population precluding the possibility of establishing long-term committed relationships. Yet, many of the single migrants I knew had already lived in Dubai for four years or more, so were not transient in their own lives, and these wider discourses were certainly challenged by some couples who had met and later married in Dubai. Nonetheless, the dominance of these discourses about transience was such that many single Britons claimed to be strategically avoiding emotional intimacy while in Dubai. For instance, my field notes draw attention to occasions when men and women closed off the possibility of establishing relationships due to their perceptions of the city:

> Nick is twenty-five and was born in Cornwall. He moved to Dubai after travelling for six months through Europe and Morocco. When I first interviewed him he had lived in Dubai for three months; it was his first experience of actually living abroad. Nick is employed as a Project Manager in the construction industry, a job he applied for once in Dubai, and he has a three-year contract. Nick met his girlfriend Rachel, twenty-five, a teacher from Australia, when he arrived in Dubai, but he told me: 'I still keep things casual because, after all, you're in Dubai. How realistic is it to expect it to last?' (field notes).

British men and women discussed the 'superficiality' they perceived in their single contemporaries of the opposite sex:

> Vicky is twenty-eight and from London. She is employed as a flight attendant, lives in an apartment on Sheikh Zayed Road, and is intending to stay single while she lives in Dubai. She told me: 'You don't wanna go there! Like, really, there's no point unless you just want a shag [sex] because they're [men] never going to be up for a proper relationship' (field notes).

For Ed, whom we met at the start of the chapter, single British women in Dubai were to be ridiculed, yet were also a temptation while living away from his girlfriend in the UK. He told me he thought Dubai was conducive to being unfaithful, 'with all the trolley dollies [derogatory term for female flight crew] and the holiday attitude that girls bring with them'. He put on a high-pitched voice to mimic a woman: 'get burnt, get drunk, and shag some bloke!' (field notes). Similarly, Malcolm, a primary school teacher from Newcastle who had lived in Dubai for three years, felt that staying longer in Dubai was risky because he perceived the single British women he had met as unsuitable for committed partnership: 'most of them are slappers', he told me (field notes). However, British women in Dubai were equally critical of single British men they observed. Jenny was twenty-nine and had been working for two years as a fitness instructor in a five-star hotel, living in Al Garhoud in an apartment provided by the hotel. She explained her single status in relation to the lack of eligible men: 'Men my age, and even older, get away with behaving like twenty-year-olds out here. They're with a different girl each night or they're paying for prostitutes' (field notes).

Giddens (1992: 69) remains sceptical of women being able to organise their sexuality as a man might 'without too many psychic problems' and draws on therapeutic discourses to suggest that women's 'promiscuity' results from sexual addiction, self-disgust and loss of self. I would argue against this gendered interpretation, since it takes away any agency from women to engage with heterosexuality through the fun and spontaneity I observed. However, I am also cautious about suggesting that single straight femininities in this context can be interpreted in relation to progressive and empowering 'new femininities' (Laurie et al. 1999), since the emotional geographies of these textures of intimacy suggested they were experienced with more ambivalence. Likewise, the masculinities in evidence were certainly not transformed by the wider currency of the 'new man'. It is necessary to explore, therefore, how gender inequalities are reproduced in complex ways through the cultures of sexuality and romance that emerge in transnational urban spaces. While both British men and women in Dubai enacted and were critical of 'promiscuity', the scripts of sexuality and romance available to them in constructing their intimate subjectivities remained distinctly gendered.

In particular, the heterosexual subjectivities produced in Dubai that I observed were gendered through discourses of emotion. In talking about themselves and each other, single British expatriate men and women frequently drew on the gendered stereotypes that are prevalent in popular therapeutic discourses and widely reproduced in self-help books and the media. In such representations, heterosexual relationships are constructed through a series of dualisms: femininity/masculinity, emotional/unemotional, irrationality/rationality, attachment/detachment and dependent/independent. The label of 'girl' that Jenny used also reminds us that expatriate women are located within a culture of romance, with particular notions of femininity at its core (Redman 2002). In contrast, single British migrant men who adopt the practices and performances of a more widespread 'player' masculinity become empowered through the idea of relationships being a 'game' and women a 'conquest'. This way of thinking remains remarkably persistent in discussions of couple relationships, despite an increasing acceptance of men's emotions in particular contexts (Van Hoven & Hörschelmann 2005). So, while some women laughed as they told me they 'went a bit mad' in reference to a higher number or frequency of sexual partners in Dubai, discourses of mental (un)health are also used, by women and their friends but also by men, to describe their approach to relationships, particularly their confusion regarding its actual (or potential) level of intimacy. Men were often very disparaging of women who made an association between sex and the potential for a romantic relationship. These field notes reveal one instance:

Mark is thirty-three, from London, and has lived in Dubai for five years. He was complaining about a woman ringing him too often. He said he couldn't believe how he always managed to pick 'the clingy ones' out of a room full of sexy women. They [his friends] discussed how he should ignore her until she gets the message that he doesn't want to have a relationship (field notes).

Equally problematic, however, is the way in which expatriate masculinities are often portrayed in reductive ways by British women through the construction of men as unemotional. Women seem to understand this construction either as a deliberate (sometimes cruel and manipulative) detachment or as a more benign, yet inherent, incapacity to be in touch with their emotions. When I met Tanya for the first time, she explained: 'Everyone's got baggage! When you meet a guy out here they tell you they came for the job or the sunshine. It may take a while but eventually you find the real reason. Suddenly it dawns on you and you sit there thinking: 'That's why you're in Dubai!' (field notes).

Although social theorists (e.g. Giddens 1992) have argued that our emotions have become necessary props in the creation of subjectivity in the contemporary age of psychological and therapeutic knowledge, displaying emotion, or not, remains a crucially significant way of performing gender. As Whitehead (2002: 156) argues, 'there is a commonly held view in societies that men "cannot do" relationships as effectively as women' because masculinity 'falls short when it comes to facilitating and enabling the emotional labour required to sustain a relationship'. So, it is not that single British men do not experience mobility in emotional ways, nor that women are inherently 'feeling more' in the transnational context but, rather, that Stevi Jackson's (1999: 108) argument about young people can be extended to single men and women in certain contexts: 'Being constituted as feminine involves girls in discourses of feeling and emotion and, more specifically, the culture of romance, from which boys are excluded or from which they exclude themselves in order to construct a sense of their own maleness.'

British migrants also drew upon cultural resources from the Western media to position themselves, in gendered ways, within a set of discourses about love, sex and romance as part of a making sense of the textures of intimacy in Dubai they encountered. In doing so, they constructed their emotional subjectivities through notions of intimacy. For single women, this often involved identifying with the scripts of dating that circulate in song lyrics, international television shows, and self-help books, in which surveillance of relationships and responsibility for how they function are distinctly gendered (Reynolds 2008). During my fieldwork, *The Complete Book of Rules: Time-Tested Secrets for Capturing Mr Right* was one such resource; another was the popular television series *Sex and the City*. Although some migrants were critical of their messages, most of us devoured them with relative passively. Indeed, Akass and McCabe's (2004) edited collection of essays on *Sex and the City* convincingly argues for the cultural significance of the television show in redefining the single woman and her attitude towards dating practices. Such gendered discourses then emerged in localised forms in Dubai's expatriate media, as evidenced in an article called 'Disappearing boyfriends' that appeared in the newspaper *7Days* and in which the author argued: 'You see now-adays, it's easier to see a guy and sleep with him, than call him your boyfriend... Relationships are minefields of emotional drama, where the only thing that keeps us steady is a glass of Chardonnay' (Knight 2003: 12). For Reynolds (2008: 151), the marginalisation of single women in contemporary British society and the con-tradictory nature of cultural resources mean that there is 'difficulty in performing a "single" identity that gives an empowered sense of self'.

Conclusion

The significance of heterosexualities is under-examined in literatures on migra-tion (Walsh, Shen & Willis 2008). This is also the case in literatures on British migration, where the heteronormativity of 'expatriate' subjectivities has escaped critical attention. Literatures on Gulf migrant subjectivities have also been silent on this aspect of migrants' subjectivities, with the notable exception of Mahdavi's work (2011, 2016). Yet this chapter, in exploring the heterosexualities of single British migrants, has demonstrated how intimate subjectivities are reconfigured in relation to changing ideas about romance, love and sex. The empirical material suggests that Dubai itself, as a specific site in which these subjectivities are enacted, is extremely significant: the nightscapes of the global city shape the pro-duction of migrant subjectivities described as classed, racialised and gendered in particular ways. For example, the gendered negotiation of Dubai's 'decency laws' constrains migrant women's heterosexualities and reproductive rights by making not only sex, but also pregnancy, illegal outside marriage. Furthermore, the depiction of Dubai as both a 'holiday-like' and 'transient' space influences men's and women's experiences of migrant lifestyles, with the notion of 'escape' being positioned against those of domesticity and heteronormative expressions of heterosexuality.

The evidence suggests that single British migrants understood Dubai as a space of liminality, where the heterosexualities deemed appropriate emphasise sexual desire over romantic commitment. Yet, the pervasive ambivalence about couple intimacy evident in 'expatriate' talk suggests that notions of romantic love and marriage had not lost salience for the many single British migrants among my interviewees. While many embrace a lifestyle free from obligation or commit-ment, so that it might be tempting to portray them as the archetypal figure of 'individualisation' (Beck & Beck-Gernsheim 1995) in a globalising world, priori-tising their career trajectories and travel adventures over the 'stability' of love relationships, in fact this lifestyle is usually temporary. Berlant (2000: 2) high-lights the contemporary strength of 'the modern mass-mediated sense of inti-macy,' arguing that it leads 'people to identify having a life with having an intimate life'. As such, it is no surprise that many of my interlocutors either established longer-term relationships in Dubai or did so later on, after returning to the UK.

Marriage and children quickly followed, something which reinforces the narratives of this chapter: Dubai was understood as a holiday-like space where new intimate subjectivities could emerge and be 'tried out', but the performances of heterosexuality were often as transient as the migrants enacting them.

Migration can be understood as a self-reflexive project in which new subjectivities emerge, including intimate subjectivities revolving around sexuality (e.g. Gorman-Murray 2007; Mahdavi 2016). International relocation is already established in the wider literatures as enabling queer practices, identities and subjectivities to emerge (Manalansan 2006: 225) and this seems to be equally applicable to non-normative straight subjectivities. Indeed, for Wilkinson (2014), the 'single' is a potentially queer subject, since space normalises coupledom. This is especially the case in Dubai, where heteronormativity is inscribed by both the dominant lifestyle among the British migrant community where marriage remains the norm, as well as by the intimacy codes encoded in the UAE's decency laws. Like all sexualities, then, British migrant heterosexualities involve 'continually negotiative enactments in relation to public discourses and interactions' (Walsh, Shen & Willis 2008: 576), and are reconfigured in the global city. Of course, heteronormativity regulates not only those whose lifestyles contest heteronormative expressions of intimacy, but also those who enact the norm. As such, the next chapter takes a closer look at British migrants' family lives, exploring how familial and spousal relationships are also shaped by migration to Dubai.

Chapter Seven
Migration, Domesticity and 'Family Life'

As Valentine (2008) notes, families have been neglected in geographical studies, in spite of considerable attention given to childhood and parenting. Yet, notions of heteronormative family life are closely wrapped up with the space of home (e.g. Dowling & Power 2012) and, therefore, emerged quickly from my study, given the initial methodological focus on domestic materialities. Indeed, ideologies of home have familial relationships, parenting and childhood at their heart but, as such, they also reproduce gendered subjectivities through heteronormative expectations of care and intimacy (Bowlby, Gregory & McKie 1993: 344). Consequently, this chapter has the reproduction of intimate subjectivities through the negotiation of home as its analytical focus. First, the chapter draws attention to the significance of 'family' in the motivation for the economic/lifestyle-oriented migration of this group and the constitution of family through practices of togetherness. It then moves on to consider how the heteronormative ideal of feminised familial care is challenged by women's resistance to the division of work/home and associated performances of femininity and masculinity on which it is based. Both these discussions highlight how families are embedded in the broader political economy of global work, such that the making of family is far from a private concern. The third and fourth sections consider the significance of relationships with family members who do not share a domestic space for the adult migrants' sense of home and belonging. The third section explores family display and the making of a family home for adult children to visit, as well as the dismantling of a family home in the UK. The final section discusses the transnational connections, especially to ageing parents, connecting with wider literatures on

Transnational Geographies of the Heart: Intimate Subjectivities in a Globalising City, First Edition. Katie Walsh.
© 2018 John Wiley & Sons Ltd. Published 2018 by John Wiley & Sons Ltd.

the circulation of care and visits between family members in different countries, especially in relation to ageing (e.g. Baldassar & Merla 2014; Lamb 2002). These sections suggest that families are constituted through family practices, including display, visits, and gift exchange, in spite of separation in everyday life. These migrants are, therefore, part of the variety of transnational family forms that Baldassar and Merla (2014: 6) point to when they argue for the 'need to theorise mobility and absence as common features of family life'.

'Family Time' and Space for Family

While it is now well-established that women accompanying their husbands on international assignments often find it impossible to continue with their own career, due to visa restrictions, lack of suitable positions, language barriers, invalidity of their qualifications in another country or the impermanence associated with mobility (Hardill 1998; see also Coles & Fechter 2008), this was not always considered a problem by the British women I met in Dubai. Many women were not career oriented prior to their relocation and so they valued an opportunity to become 're-domesticated' and give up their jobs. In extreme contrast to the 'migrant mothers' whose family lives are marginalised by their employment as transnational domestic workers (e.g. Pratt 2012), British migrant mothers may instead experience an increase in the time available to them for family life. The relocation to Dubai enabled them to spend more time with their children than they would have been able to afford to do at home. These mothers also appreciated being able to reclaim time to spend on themselves. Therefore, many women considered the relocation as something that enabled them to be more effective in gendered roles of their choosing: to perform 'better' as wives and mothers. Field notes, taken during a conversation with Claire, reveal both these sentiments:

> I absolutely love it! I hated working. I hated commuting. I used to sit on the bus on the way to work and count down the days to the weekend. And then I'd have to shop on the way back, get home, cook supper, do the ironing. I feel blessed that out here I don't have to do that, that salaries are high enough, you know? I don't know how I'll go back to working if we move back. I can't think why anyone would want to work if they really knew they didn't have to. I have great friends and we have a great time. I swim three times a week. Dom plays golf on a Thursday, but it's the children's weekend so I feel I should be there. So [if she were in paid employment], I just wouldn't get the 'me-time'. We have girls' nights out most weeks too, which my friends have maybe once a month, if that, back home. I think it makes you a better person, a happier person. I like being a Jumeirah Jane for a while. Why not? Most people would if they got the chance. My sisters are jealous when they come out here, because they're on holiday and I'm just living my normal life. I'm not going to get some crappy part-time job for a few Dirhams an hour when there's plenty to do. I'm healthier than I've ever been. I look better…I make a better wife (field notes).

Claire highlighted both her own lack of paid employment and the replacement of her domestic labour by someone else (a domestic worker) as creating time. Time is important in family life since family is constituted through doing things together (Morgan 1996: 2013), yet her narrative also reveals how more time allows family members to do separate things, alone or with friends. This shows how, for Claire, migration becomes framed as a comparative project, something familiar in analyses of lifestyle migration (see also Benson & O'Reilly 2009). However, Claire's celebration of her lifestyle in Dubai highlights economic changes as a route to altered everyday routines and the consequent shift in quality of her personal relationships with her children, husband and friends.

Claire was one of the few women who told me such a wholly positive story, but the narratives of working women also pointed to the strength of commitment from many women towards more conventional gender roles and the homogeneity among the 'expatriate community' in this respect. My interview with Jenny demonstrated this, yet also revealed her resistance to these dominant views and practices of migrant motherhood:

> I am judged by a lot of the non-working mums here as being an inadequate mother. I have actually had two women on separate occasions tell me that they don't know how I could leave my son to be reared by other people, when it is my responsibility. I really question how much time they actually spend with their kids, between their hectic social lives. Most children I see here are at the park with their maids and most women I see shopping also have their maids with them to mind the kids. I take my son to dinner, to lunch and shopping, to the park, the pool, and even to the soft play areas. When I am not at work I am with my son and husband. Yet, I was never so openly judged by women at home, as most women do work and raise children. These tasks are not done in isolation of each other: it's more like they complement each other. I feel like this is how the women who wanted to work in the fifties must have been treated (interview).

Claire and Jenny's accounts are two of the stronger views that were expressed but, as this chapter will show, most non-working women were ambivalent, identifying both advantages and disadvantages to the shift in motherhood practices they experienced in Dubai.

One of the ways in which these shifting negotiations of parenting and family time are materialised in Dubai is in the creation of 'family rooms'. In British houses it is rare to have a 'landing' at the top of the stairs large enough to furnish more than perhaps a bookcase, storage box, or computer table. In Dubai, however, many of the large older villas incorporate a room-sized space at the top of the stairs. Some of my interviewees seemed at a loss as to what to do with this 'extra' space, leaving it empty or using it for storage (see also Thomas 1999 on Vietnamese migrants adapting to Australian architecture). Yet, in other homes I visited, this had become one of the most used spaces in the house, as an informal living area. The connections between the domestic home space and family are well established. As Chapman (2001: 144) argued:

> at a surface level, home is known in terms of its location, fabric, decoration, furnishing and amenity – it is a place we know intimately. At a deeper level, home is defined in terms of the kinds of relationships people have with others in the home, or [...] the kinds of relationships they would like to have.

More recently, David Morgan (2013), writing about 'doing family', argues that the home is a key site for the doing of family practices through which family is constituted and family life unfolds. And based on their empirical work in Australia, Dowling and Power (2012: 608) demonstrate empirically how a house is made into a family home 'through providing spaces and opportunities for members of the family to be together, to relax and feel secure in their shared living space'. These perspectives resonate with the way in which some interviewees understood this new domestic space they found in their Dubai villa. Lucy explained that their 'family room' at the top of the stairs had become the room in which the family spent time together:

> This is where we live, where we spend our casual time. I think they are made like this because if guests come to the door the Arab women come upstairs to hide until they know whether the visitors are male. We keep all our hundreds and hundreds of books here. Everyone reads in our family. There was a TV point already and we put the computer up here as well, so we're together even if somebody's on the computer, somebody's watching TV, somebody's reading or whatever. It's the focal point. It's stupid really because in the evening the bedrooms are right here so you get the noise of the TV or whatever, although in the summer it's fine because you have the noise of the air con. But, actually, I would prefer to have a lounge where you can shut the door – it's too open. It has worked though, because we've got the work-station, everything's here. That's worked, that we actually spend a lot of time together, but I do like the idea of having a lounge where you can actually shut the door. You know, sometimes, if Carrie is playing the piano, it's all you can hear, you hear it whether you want to or not. So when she's doing scales for half an hour you think, 'Oh god! Shut up!' which isn't fair because she has to practise. So, for that, it'd be nice to have a door you could shut. But the design of them, you'll see that they're all like this (interview).

Dowling and Power's (2012) research in Australia demonstrates the allure of larger home spaces for contemporary Australian families. Their interviewees had chosen to live in large detached suburban houses which enabled them to organise the spatial and temporal routines of everyday domestic life in ways deemed appropriate to performing respectable middle-class family life, parenting and childhood. With larger and multiple living spaces, as well as bedrooms for each child, the parents chose to contain children's excess, with the 'messiness' of their toys and noise confined to a playroom, as well as exclude them from adult-only zones, whether in the form of retreats for the parents or formal lounges for adult guests. Lucy and Matt's room upstairs was also a private space within their villa, allowing rooms downstairs to be designated as more formal spaces for entertaining guests. Thereby, the organisation of the rooms in this way neatly avoids compromising the family's comfort when their home is frequented by visitors. On the morning I visited, for example, there was a half-finished bottle of orange fizzy drink on the coffee table, some DVDs scattered on the side, an unread paper, masking tape and folded plastic bag lying on the desk, an adapter plug stored on the bookshelf, and a wastepaper basket full of rubbish. It was far messier and more cluttered than other areas of the house. Not intended for public view, their family room was also more personalised through display; for instance, a rugby shirt displayed on the wall, pictures of family in frames on the chest of drawers, a large collection of books, and a little orange paper flower stuck on the corner of the television. Knowing Lucy's experience of friendship in Dubai was one of relatively superficial friendships (see Chapter 5), the decision to create this family room might be understood as a way of removing family life from the surveillance of colleagues and acquaintances visiting their home with whom she does not feel intimate. However, it must also reflect Lucy's desire to integrate her family against the centrifugal force that results from her teenage daughters increasingly wanting to spend the evening involved in doing different things, as well as her husband bringing work home with him to do in the evenings. Her prominent collection of pigs, each one a present from Matt following a business trip, can be read as further evidence of the challenges of her role as family-maker.

The significance to both parents and children of spending time together in the home is made evident by Jeni Harden et al.'s (2013) study of the impact of working parenthood on the everyday lives of children of primary school age. Having gathered both children's and parent's perspectives, they suggest that 'home' was viewed as a 'special location of childhood', especially as a site of parental-child interaction and 'the place that seems to hold meaning as the backdrop to family life'. They noted that parents 'felt guilty about their work taking them away from being at home with their children' and the importance of being at home with their parents after school was consistent in children's accounts (ibid.: 302). Christensen et al. (2000: 146), who worked with 10–12 year olds in schools, also comment on the 'very ordinariness and non-eventful quality of time at home in the family'. They explain that some children, 'found it very difficult to articulate in conversation with us exactly how they spend time with their family... They would struggle to tell us what they actually did together...family time at home simply is' (Christenson et al. 2000: 146). However, Lucy's narrative reveals the gendered nature of family time in many households in Dubai:

> I'm the one who's at home more than he is. He's out working all day: he doesn't spend that much time here. When he does spend time here, then he's with the kids or on the computer, or we're out doing things, like diving, so he doesn't actually spend that much time at home (interview).

Lucy's frustration with this situation was evident from her tone. In the next section I explore the impact of global work on British migrants' negotiation of their marital and family lives more generally.

Global Work, Expatriation and Marriage

In Chapter 1, I drew attention to a set of literature within the social sciences that suggests that (highly) skilled labour migration is strongly gendered, with men being significantly more likely to become the 'lead' migrant through their employment trajectory (e.g. Coles & Fechter 2008; Hardill 2002). Indeed, Kofman and Raghuram (2005) observe that research on couples participating in flows of skilled migration has consistently identified the persistence of more traditionally gendered divisions of labour within households, irrespective of nationality. While many women I interviewed had already experienced this gendered 'expatriate' culture in previous postings, they also argued that moving to Dubai had been surprisingly challenging due to factors that included very long working hours (except for public sector employees who might have split shifts or shorter working days), higher expectations of travel (Dubai is often used as a base for those who need to travel throughout the Gulf, the Middle East, or more widely) or a six-day working week. In addition, many British men were also removed from domestic life during the weekend, since they participated in leisure activities for relaxation or networking opportunities (e.g. business group meetings or playing golf). For couples and families, this working culture limited the time that they spent all together. This issue was raised by both men and women, and by those who were comparing their experiences to the UK or to expatriate life elsewhere. In the majority of cases, longer hours were linked to an increase in responsibility and would still be comparable to those of many professionals working in the UK, but they were experienced as linked to the relocation. Likewise, although this situation might actually be emphasised for couples who are both working full-time hours, it tended to be an issue raised by non-working women. Therefore, it might not be the working culture per se that creates most sadness or frustration but, rather, that the gendered characteristics of this working culture mean that non-working women feel their husbands are being removed from domestic life precisely when they themselves are spending more time at home.

Certainly, it seems that British men's masculinity became figured by British women in terms of absence from domestic life. An orientation advisor I interviewed explained the experience of many women:

> You wake up at the crack of dawn when your husband leaves for work and then time stretches out in front of you, empty, until you're still waiting twelve hours later for him to get back. It's too hot to go out, and you don't know anyone, and you don't have anywhere to go (interview).

Individual narratives reiterated this representation of men's absence from home, as this example from an interview with Kate demonstrates:

> [the maid will] be the one who gives the children medicine: my husband wouldn't even know they were sick. My husband works four days when I don't see him, then he has one day sleeping and three days off, so his shifts are very bizarre. Husbands either travel a lot or are not contactable and therefore the wives become very community related (interview).

The significance of VCOs such as the Dubai Adventure Mums, as well as friend-ships among non-working women, is clearly magnified in such a context, as dis-cussed in Chapter 5. Yet wider discourses about British migrant homemaking in the city do not take the difficulties of such women into account, more often affirming the high status afforded to British masculinity through men's participa-tion in paid employment. For example, the British editor of a magazine about home interiors commented: 'The women are sat at home with big open spaces. They want to know: What they can do with their cash? How do they fill their lives?' (interview); while, in contrast, the British manager of a relocation company suggested:

> If you've got a guy working for a bank and he's earning a lot of money for that company on a daily basis, the last thing that company want is for him to be worrying about the housing and 'Is the electricity being switched on?' and 'What about my dogs, they're arriving today?' They want him in the office. They want him to literally hit the ground running (interview).

These quotes are illustrative of the way in which the 'triviality' of (feminised) domestic work is frequently juxtaposed with the taken-for-granted 'importance' of the (masculinised) employment of the lead-migrant. The cultural resilience of gendered practices of homemaking in heterosexual households (Chapman 2001) is clearly reproduced by these migration processes and discourses which reinforce the conflation of high status with masculinity. The knowledge and skills involved in the kinds of professional jobs most married British men accompanied by non-working wives are employed to do in Dubai are highly rewarded in the global economy and associated with masculinity through notions of organisational power, management and status. In spite of unprecedented numbers of female Britons now working in highly skilled jobs in global cities, these were still relatively rare in Dubai at the time of my fieldwork, and the conflation of masculinity with professional, managerial or technical expertise was amplified in this British migrant community where there are relatively few Emirati nationals in the work-place. In everyday conversation outside work, British men often undermine the status of those Emirati men they do encounter by attributing their employment to nationalisation policies rather than skill or qualification.

The transformation I observed in gendered subjectivities, towards more 'traditional' stereotypical roles, was made all the more visible because it goes against recent trends in the UK and many other countries. Men's greater involvement in the domestic work of home has 'challenged the home as a site of femininity only, creating masculine models of domesticity' (Gorman-Murray 2008: 371–372). In migrant households where women do manage to maintain their own career, a more egalitarian division of household labour is common and homemaking is sustained as a collaborative project. For example, Becky (late twenties) was a journalist by profession and though she accompanied her husband (who sought employment prior to their move and would, therefore, be categorised as the 'lead migrant'), she continued to work full-time in Dubai until the couple had their first child. Therefore, their household division of labour did not change immediately with relocation, but later with the impact of motherhood, something which took Becky by surprise: 'Now, with Christopher, it's harder because your relationship shifts. I'm at home a lot, he's at work... My car's broken this week so I'm thinking "Oh my god, I'm trapped like an Emirati lady in the house!"' (interview). This couple continued to resist the 'traditional' division of labour dominant among British migrants in Dubai, with Josh instead taking a more active fathering role in his time at home.

Janet, whom we met previously spending time with her children at the Sailing Club while her husband was away for work (see Chapter 5), provides a useful account of love, marriage and migration through which to further explore these gendered dynamics of British households in Dubai. Janet is married to Graham; they are both Scottish and in their mid-thirties. They have two children living with them in Dubai: Jane (eleven) and Ben (seven). Graham is employed in Dubai as an engineer, while Janet does part-time volunteer work. The family lived in Scotland for two years before moving to Dubai but they lived in Nigeria for five years when the children were small. For Janet, relocating to Dubai was an easy decision to make in order to prioritise their nuclear family:

> Graham was travelling a lot so he was always away when we were living in Scotland. When he got this job, I could see us never being together except on his vacations, whereas if we moved, I'd be away from my family at home, but the children would get to see their father and we could do things as a family (interview).

Their home displayed the significance of this love relationship and marriage in Janet's life. On the upstairs landing were framed photographs of the couple on their wedding day and also a painting, to which Janet brought my attention during a tour:

> That's the church at home where we got married. I was in a gallery and I just spotted it and it wasn't long after I got married and I thought 'Oh, this is the church I got married in ... It's just a little church, in a little village; in fact, it's where I used to go to school as well (interview).

However, Janet also revealed the strains that their relocation to Dubai was having on her marriage. She described the Dubai Adventure Mums and the Sailing Club as vital forms of social support for her, as Graham was so often expected to travel away from Dubai for work. The Sailing Club offered her a semi-public space in which, in the absence of her husband, she could socialise alongside other adults while her children played (so she often took them there after school to have their evening meal in the bar). Janet explained that she had recently been very upset at spending so much time 'alone' and had been arguing with her husband over the cost that was incurred in such activities. He had seen them as unnecessary, while Janet considered them to be essential to counteract her isolation. Janet's account resonates with that of many other British women I met in Dubai, who felt that they were taking on additional responsibilities for the making of family and home during their residence.

With over twenty years of living abroad as a non-working 'expat wife', Eve also had considerable experience of the gendered expectations of such a role. As a graduate, who met her husband at university, Eve had high expectations of her marital relationship being egalitarian, but explained to me the difficulties of making international careers work for a family:

> It's all very well when the family's going well, but we all need to escape sometimes. It became dangerous to remain living in Pakistan so we had a complicated move to here. I was the family manager: organising our time in the UK, travelling, and here [Dubai]. It was stultifying. Eventually, I just got to the point where I thought, 'Who's going to take care of me?' It's very difficult to be supportive. It puts your marriage through a lot to do these overseas things. The rewards are high, but there are great sacrifices. Like, I haven't been able to have a career. How can you when you're following someone around the world? I don't really mind. You could say that I'm lucky to spend more time with my kids but I have learnt that I do need to do some work. But, at least now, we're established and happy (interview).

Eve perceived her husband's work commitments as creating a conflict with her own need for him to contribute more to family life. For Hochschild (2003), particularly in our closest relationships, we are constantly managing our emotions to feel and express what we want, expect, or think we ought, to feel. When the distribution of power and authority is unequal, the process of managing emotions is also unequal: 'the larger economic inequality is filtered into the intimate daily exchanges between wife and husband' (Hochschild 2003: 169).

British women are often resistant to the 'new' masculinities performed by their husbands, particularly when these are 'brought into' the shared space of home and start to reconfigure how domestic practices are gendered. For example, Emma, a mother of two children of primary school age, previously worked full-time as teacher in the UK and her account illustrated this shift by which masculinity becomes contested. Emma confided in her friends at a coffee morning that she and her husband Mark were arguing more frequently about domestic issues and that she thought he was becoming a bad role model for her son because of his increasingly 'sexist attitudes'. When Mark's career in pharmaceuticals gave the couple the opportunity to travel abroad, Emma had stopped teaching, embracing the chance to travel, relax and focus on her family. However, she increasingly viewed the ambiguity surrounding their new household roles as contributing to the new tensions present in her marriage, as these field notes demonstrate:

> I was at the Dubai Adventure Mums and Emma was saying, 'It used to be me having the more tiring and stressful day, so Mark probably used to do more of the house-work and he'd wash up and cook. The other day, I told him I'd had a really tiring day but he just went and watched TV. He assumed that I'd cooked. So I was shouting at him and the kids are listening! It would never have been like this in England, but sometimes he gets into his expat mode now and he'll say stuff like "I think it's advis-able" or he'll just make the decision without consulting me. I'll say, because we went to school together... "I know where you come from!"' Janet then argued that it's the general attitude of the men in Dubai: 'The power goes to their head and they begin to treat you like the women at work.' Emma agreed, 'It's Dubai.' And her friends nod their heads knowingly. I asked 'Why do you think it's Dubai?' But Emma joked that the men 'all think they're God's gift'. Janet linked it to the fact that they're building a new city and they have a lot of responsibility at work. 'They forget they're hus-bands too.' But Emma thought it was because women are considered 'second-rate citizens' in Dubai because of the local culture (field notes).

Intriguingly, although Emma had not participated in paid employment in any of the countries that she and her husband Mark had lived in since leaving the UK in the early 1990s, she explicitly linked her increasing discomfort with being 'just a housewife' to their residence in Dubai specifically. Emma was not unique in doing so and her identification of the influence of place-specific challenges to the negotiation of marital intimacy was supported by the observations of a counsellor working among the 'Western expatriate' community more generally. The counsel-lor's interpretation was that it was often the insecurities generated by the con-tinual frustration of British men dealing with Emirati culture in the workplace that led to verbal put-downs in the home. Worryingly, she argued:

> Many women feel helpless in this situation because they do not have a lot of options. They think to themselves, 'Do I return to the UK and live in poverty or do I hang on in there?' The men here would say they put their women on pedestals, but I would disagree (interview).

This counsellor reported a high incidence of 'situational depression' among accompanying spouses, a term she explained would be applied to those experi-encing adjustment difficulties arising from the trauma of recent life events, in this case relocation. This term seemed to resonate with the experience of Frances who gave me a photocopy of a poem by Barbara Bassett called 'Angela's Word', taken from the popular self-help book *Chicken Soup for the Woman's Soul* (Canfield et al. 1996), to help me understand her life. Frances had been a British resident in Dubai for over twenty years and was open with me about her recent turn to counselling for support and the fact that her doctor had prescribed anti-depressants to help her cope. She did not dwell on the meaning of the poem, merely stating that it was important to her, but I believe she gave me the poem to help me understand how she felt: it described a woman struggling with her feelings of separation from her husband and herself. For some women, the emotional stress of trying to cope with the gender roles and expectations of the expatriate lifestyle can irreparably damage conjugal intimacy. While there is no statistical evidence to support their assertion, the counsellor I interviewed also claimed that divorce rates are significantly higher among British couples in Dubai in comparison with the UK population more widely, and linked this to changing or stressful gender roles (interview). The British Community Assistance Fund (BCAF) also uses the term 'abandoned wives' to describe the many British women they support with financial aid for repatriation and counselling when they become separated or divorced from their husbands in Dubai.

Adult Children and the Family Home

Domestic artefacts are significant in practices of displaying familial relations. As Janet Finch argues:

> families need to be 'displayed' as well as 'done' because their contours and character are not obvious in an environment where relationships and living arrangements are both diverse and are subject to significant fluctuations over time, even quite short periods of time. At the same time, the intimate relationships which 'families' represent are of fundamental importance to each individual's sense of place in an ever-shifting social world, which itself requires recognition by others. So relationships need to be displayed in order to have social reality, though the intensity of display will vary in different circumstances and over time. But whatever the circumstances, the core message of displaying is 'These are my family relationships, and they work' (Finch 2007: 73).

The tools for display that Finch (2007) identifies include both the domestic artefacts of everyday life, including photographs and inherited 'keepsakes', as well as the use of narratives, stories about family life through which the character of family can be communicated to a listener. Of course, the listener might include the researcher and certainly it was a central part of my analysis, right from the beginning, to consider how individuals, couples and families were displaying their relationships to me as they showed me around their homes. Finch (2007) highlights that the degree of intensity of family display varies with circumstance and it is perhaps not surprising, therefore, that it was often most evident in interviews with women whose adult children no longer lived with them. It is widely recognised that the parent–child bond, as well as the parenting practices which help constitute family, do not automatically finish when a child reaches adulthood. For many British couples in Dubai with grown-up children, the decision to migrate was shaped by their feelings towards their children and sense of responsibility and connection. While experiences are diverse, here I focus on two couples whose narratives reveal how this might also be shaped by gender, privilege, career trajectories and previous experiences of migration.

First, the narrative of Amanda is helpful to explore. Amanda and Bill had lived in Dubai for approximately five years when I conducted my interviews, relocating from Cheltenham, Gloucestershire, in their early fifties. The couple had a much longer history of expatriate travel since their marriage, having lived in Tehran, Singapore and Hong Kong while raising their children. Two of their children lived in the UK while the third worked for an international bank and was himself posted to a new country every six to twelve months. Amanda's parents were also expatriates so, as a child, she went to boarding school in England and stayed with her grandmother in Wales during her vacations. With his extensive experience of working overseas, Bill had attracted a request from his company to open a new overseas office in Dubai and was able to negotiate a relocation package that included a large allowance for moving house and the rental of a five-bedroom villa in Umm Suqeim, an expensive and desirable residential area in Dubai. This enabled Bill and Amanda to also keep their apartment by the Thames in London and to take with them to Dubai as many of their belongings as they desired. Amanda explained her decision to take advantage of this:

> When I realised we were coming to live abroad again, I said to Bill: 'Look, I'm just not prepared to camp, you know?' [...] You've really got to live your life properly, fully, is what I'm trying to say, and if you're not comfortable in your surroundings, um, when I say comfortable, things like, we brought Granny's table and our pictures [Amanda gestures around room]. So, it's just homely, and when the children come here, they're coming home. Even though we've got a small flat in London, this is home, this is home for them.

Having described their efforts to establish a sense of their villa in Dubai as the primary family home, Amanda continued to assert the importance of recreating home upon relocating:

> It's [home is] wherever you are, do you know what I mean? Otherwise you're living in a shell, surrounding yourself with things because they're cheap. You know, you like nice things around you. So that's why we brought all our books and we made a conscious decision. All the tables are ours...the sofas, china, cutlery, beds. It's somewhere where people can come. We have dinner parties. It's very nice to be in one's own ambience. People do laugh at the British though, because they tend to think that we never really leave England, we bring our drawing rooms with us. Which I think in many ways we do, you know, because you could put this drawing room just

about anywhere in the world [Amanda laughs], couldn't you? It's very important...I always tried, subconsciously, I didn't know I was doing it, my mother used to make our bedroom the same wherever we were, so the same curtains would go up, altered if necessary, wherever we were we knew what it felt like, and I used to do that for my children as well, and it meant they settled rather quicker (interview).

Amanda demonstrated pride in her homemaking role for her children, not only for her past efforts while they were growing up, but also for her continued attempt to make a familiar space in Dubai. As Tony Chapman reminds us, 'women are neither necessarily powerless nor intrinsically dissatisfied in the domestic sphere' (2001: 144). As well as drawing on the gendered and classed traditions of her own family, through her language of the 'drawing room', Amanda ties her home-making skills into a wider history of British domesticity and the making of respectable British families abroad. Alison Blunt (1999) has shown that middle-class British women were expected to play a central role in reproducing British imperial power through their practices of household management. In so doing, feminised discourses of domesticity were assumed and reiterated, and the appropriate behaviour of women as housekeepers and wives was defined. In contemporary British homes in Dubai, as this chapter has already demonstrated, it is also predominantly women who are responsible for the materialities of the domestic sphere and, consequently, shape the emotional continuity and familiarity of home for their children, however adult or infrequently present.

Furthermore, some objects were particularly important in constructing and communicating this familial sense of belonging. Amanda told me unprompted about a picture on the wall of the living room:

Although I went to school in England, that was possibly the only time that I really did live there. I had a Granny in Wales, so I could go and stay there. It's funny because you almost become a nomad and I didn't want that to happen [to her children]. Which is exactly what we did for our children. You see that cottage there, that cottage is in Wales and that was home as far as they were concerned. Every summer they [the children] were back there and, actually, every Christmas up until the age of five, because my parents had retired not far from there; and so, you know, they always felt they were English. The little church from the cottage gate is where our daughter was christened. We had that cottage for twenty years, and that's another view of the church just round the corner [points to another picture]. The dog in the [first] picture is Shambles who we took back to Singapore. It's a lovely part of Pembrokeshire. [Katie: 'Who painted the pictures?'] He was a neighbour of my parents across the bay and he got the train over one day and painted all three, which is fantastic. We loved it down there, but it just got too busy to keep it, and we never got down there for long enough, and in the end we just had to sell it. I'd love it now. The whole circle has gone around now and it would be lovely to have it back. We always had a home in England: that little cottage. We didn't have to take anything. It was ready to go. It's very important, it's terribly important actually. I think it's awful for people who think they'll just keep themselves liquid and not buy themselves a home to go back to in the UK. I just think it is so important to have something, some sort of base (interview).

Amanda's testimony draws our attention to this painting as a representation of family history and belonging, even while she is clearly aware that the dwelling and place itself is irretrievable and mediated through memory. The house was located in Wales and only served as a holiday home while the family lived in Singapore, yet it was used by Amanda in the interview to construct the Englishness of herself and her family. This revealed not only her efforts to maintain a sense of a national homeland, but also an unintentional slippage that illustrates the inconsistencies and fluidity of the construction of home and belonging.

In contrast to Amanda's narration and clear sense of agency, Susan found the logistics of her relocation – clearing the family home and packing up their belongings – a 'horrible' experience, 'very, very complicated; very painful' (interview). Two of her three adult children were still living at home when Tom applied for a new job in Dubai, so Susan initially decided not to move with him from the Midlands to Dubai (six years previous to our interview). Instead, Tom had relocated first, taking with him only his office paperwork as air freight. Susan was intending to stay in the family home in Britain and maintain her relationship with her husband internationally until he returned. Susan explained: 'I didn't want to move area, let alone country.' Couple relationships conducted across two households have been identified in existing research and termed 'living-apart-together' (LAT) relationships (Levin 2004). However, I did not find examples of this kind of relationship among couples who were married. Indeed, when Susan became familiarised with Dubai through visiting Tom and felt that her home in the Midlands had lost meaning without their co-residence, she decided to join him more permanently. Since Tom had applied for his position in Dubai on a voluntary basis, he was not provided with a generous expatriate package. With their relocation allowance, the couple could only afford a small container to transport some of their belongings. Similarly, the salary and accommodation package his job commanded reflected the fact that they did not have young children and Susan was not expected to return to Britain each summer and, therefore, would not need a house to return to. As a result, Susan had to help her children move out and then prepare the house for rental by putting the contents into storage.

On a tour around the villa she now shared with her husband, Susan emptied one of her kitchen cupboards to show me her possessions. Unpacking the contents (glass and ceramic mixing bowls, plastic colanders, a set of scales, glass and ceramic oven dishes, baking trays, jugs and flasks, a food processor, a measuring jug, empty ice-lolly moulds, and some Tupperware containers), she explained that these were the things she had chosen to bring with her to Dubai:

I wanted my own kitchen things. I'm a very kitchen person and I like cooking, and I like to entertain, so I wanted my own things. There used to be five of us and

I always used to cook a big pot of whatever it was. Now, I have to be careful, because I still end up with the same size lasagne and we have to put it in the freezer, and we're eating it for twelve months! Now, I don't do cooking and baking like I used to. I always had homemade biscuits and cakes in the kitchen at home, but there's no one to eat them now and we're usually watching our weight, so I don't cook them anymore (interview).

In Susan's home, the kitchen objects were significant in the construction and practice of her intimate subjectivity, especially mothering. As a result, the objects reminded Susan of her distance from her adult children and her home had also become a space of loss. Transnational connections with family members and the negotiation of intimacy across distance are further explored in the next section.

Transnational Connections and Family 'Left Behind'

In spite of the economic benefits associated with British transnationalism in Dubai, migration is a matter of choice rather than necessity, a voluntary rather than a forced move. So, while British migrants are often embedded in extended kin networks in Britain, with relatives towards whom they may feel love and responsibility, often this does not seem to impact greatly upon their decision to migrate. Rather, migration appeared to be understood as an appropriate response to career opportunities or to follow a spouse in order to promote the well-being of a marriage or nuclear family. For instance, Marion's narrative demonstrated the way in which Britons think about the implications of their absence on the emotions of others, but also provides evidence of the prioritising of the couple relationship (including their lifestyle and financial security) over the needs of wider family:

My husband's father, he's 85, has just had a heart attack so it was very difficult for us to come back out this time, but we have to stay at least until after April, otherwise we'd lose the tax benefits of being here in the first place. It's difficult, but you accept that you're going to be away when you take the job, although when something like that happens it brings it home to you how far away we really are. Still, when I'm here, I think of this as home (interview).

International migration becomes possible because British norms of elderly care do not make intergenerational homes a responsibility of adult children. That is to say, most British migrants can rely on the permanence of these bonds in their lives, and a sense of their relationships being close, even when they do not live in proximity with their parents (and siblings). For these migrants, either familial love is perceived as strong enough to enable continuity in intimate relationships, irrespective of geographical distance, or intimacy is not expected from these relationships. Either way, transnational care, rather than 'hands-on' physical care, is deemed sufficient (for an extended discussion of intergenerational care see: Baldassar, Baldock & Wilding 2006; Baldassar & Merla 2014). The guilt identified by Baldassar (2008) as being experienced by many children in her research with Italian families was not apparent among participants. However, I am wary of interpreting this in terms of a distinction in British and Italian cultures of elderly care, since my methodology and my own positionality as a relatively young researcher may have discouraged the sharing of this information.

Yet, in Marion's narrative, there was also a sense that the intimacy norms expected and desired of familial relationships had been endangered by their relocation to Dubai:

> My family would prefer it if I were at home. I have three grown-up children and I miss my grandchildren very much while I'm here. Actually, I think most of all it hits my eldest grand-daughter. She's three now and I've minded her one day a week since she was born. She takes it very well really, but I've noticed, she was never really bothered about going somewhere without her parents or a family member, but now... I think she knows I'm coming back, but I think she doesn't like that feeling of missing me. I keep in touch with my family by email, text and the odd phone call, but it's not the same. My children haven't visited yet: my sons can't afford it and my daughter-in-law was pregnant and then the baby was too young. One of my sisters I'm closer to than the other, but it's difficult to find enough time to visit everyone when you're back for a holiday (interview).

As Urry (2004) suggests, people seem to have a 'compulsion to proximity' and embodied physical closeness in the practice of their intimate relationships and, as a result, engage in intermittent travel. Drawing on the work of Erving Goffman, Urry suggests that physical co-presence is understood to be crucial to many socially organised activities as it allows for emotion work, including reading the facial expressions and body language involved in communicating love. It is often assumed that British migrants are a highly mobile group with seemingly no barriers to their travel, but this is not always the case and does not take into account the varied mobility of any non-migrant family members.

Familial intimacy seems to involve not only a desire to spend time together, but also an obligation to prove your love by being present. It seems that virtual travel can only be understood as a complement to, rather than substitute for, corporeal travel (Urry: 2004). Spatial proximity continues to be desired by many, particularly around specific events that help constitute family (e.g. religious or cultural celebrations) or times of acute need (Baldassar et al. 2006), both of which exaggerate distance. Many interviewees made brief reference to their regret at not being part of the daily routines of elderly parents, but rarely dwelled upon it, for instance Mary reflected: 'just to visit, you know, have a cup of tea and check on them' (interview).

As a result, it was often disappointing for Britons to find that it was more difficult than they had anticipated to sustain familial closeness following their migration. Some perceived themselves to no longer be such a central part of the extended family networks they had 'left behind' – Emma's account of her visits back to Britain was not atypical. With her husband Mark, Emma had lived outside the UK for fifteen years and they returned annually to celebrate Christmas and the New Year with their families in Lancaster. Most years, Emma also took their children to spend the summer and Christmas holiday periods with her parents. As Emma's account demonstrates, visits 'home' are moments when the meaning of family is renegotiated:

> When we first went abroad they were quite supportive, but certainly they've never really wanted to know a lot about our life abroad. And I mean I've lived through coups, etc., but they're disappointed if you haven't followed what's happened recently in *Coronation Street*. I don't feel like we have common ground. Your life is too alien to them so they can't share it. We hide our new life totally. I think if we went back and I had conversations at home like I do here, I think they'd totally hate me! Definitely, the families Mark and I come from. This is the hardest thing, now it's Christmas. I'm buying presents and I don't know where to pitch it. You can never get it right. If I think, 'I'll buy my mum some nice jewellery because we can afford it and I'd like to treat her', everyone's, 'Well, she's got enough money', and it's not appreciated. If you are careful not to spend a lot, because you don't want to appear flashy or like you're trying to outdo people, and you buy her slippers then you know the thought is: 'With all that money, you'd think she'd buy more than slippers!' A lot of it's not said, but you just know it. And that's not your relationship that you have had with your family. I don't like it at all (interview).

While Emma expressed regret at the reduction of intimacy with her wider family, for other British residents in Dubai migration was experienced as a liberation from familial ties. Earlier in this chapter, I discussed Lucy's family room. Her husband Matt had six siblings, but described himself as closer to those who had also emigrated. He did not visit his parents or siblings who 'still' lived in Essex where he grew up. Matt called one of his brothers 'Billy Britain' (field notes) and claimed that his brother would like nothing better than to settle down in a local Travelodge, an evocatively (and ironically named) parochial image of constrained mobility. In contrast, he described himself as an 'in for a penny, in for a pound kind of guy', and considered himself lucky to have escaped his 'boring family'. Such narratives suggest we should not assume that extended familial relationships are characterised by love, nor romanticise love as being only experienced in non-ambivalent ways. However, they also remind us of Smart's (2007) observations of the 'sticky' nature of blood relations, their 'haunting' quality, in that they can continue to influence us at an emotional level even outside of physical co-presence. Matt continued to construct a narrative of his own identity through a comparison with the sibling towards whom he felt different, revealing his continued embeddedness in these more extensive familial relations, irrespective of practices of transnational visits.

It is important, too, not to forget that British migrants who move as individuals may also see themselves as part of families, in spite of not living with other family members. Jane's discussion of the decoration of her room revealed this sense of embeddedness in family life, even as an adult away from 'home', since she drew attention to the important relationships in her life:

> What I tend to do is, I have the same photos that I take everywhere, whenever I go away or move house. Usually I put them on a big pin-board or on the wall. I sort of started this off [Jane gestures to the mirror], and then it was just falling to bits and whatever [Jane gestures to the piles of fallen photographs now on the table together with those which haven't been put up yet], so I didn't carry on. They're just different places I've been, different places I've lived really. Friends and family: that's my mum and dad, me best mate Claire [Jane points to particular photographs]. It's just nice 'cos it's, like, places that I used to live. That's the view that I used to have outside where I lived in Bondi. I used to, like, walk outside my bedroom window and that was what I saw every morning. So, that's quite nice because it's [her voice gets very quiet] where I used to live. Things like that, you know, the people I lived with in Australia, and that was, like, she done a big surprise 25th birthday party for me. I've stuck them up so many times they all fall off. Or those over there [Jane gestures to the table again] are sticking together now.

The photographs were both reassuring and disquieting, reminding Jane of her geographical distance even while they evoked continued emotional intimacy. The collage revealed the way in which families are practised across transnational space in both visits and daily practices of communication and display.

Conclusion

Families should not be privileged or romanticised in a critical exploration of intimacy. Nonetheless, the salience of family life in the reproduction of British migrants' intimate subjectivities makes it important to pay attention to the renegotiation of both 'the family' (as a social construct) and everyday family life in transnational space. The term 'families' here, then, is one that encompasses co-resident households in Dubai formed by married couples travelling with their children, but also the transnational families of married and single migrants who sustain connections with parents and adult children across international borders. Indeed, a focus on the negotiation of families and family life in this chapter has revealed the significance of relationships with 'family', both present and absent. The spatialisation of intimacy is clearly not straightforward: challenges were evident in the renegotiation of intimate couple relations in transnational spaces in which gendered subjectivities were being reconfigured in relation to work and home. In this sense, the chapter has demonstrated the emotional and practical work involved in the making of heteronormative homes. While theorists rightly assert that intimacy does not necessitate physical co-presence, the processes of 'redomestication' (Hardill 1998) that British migrant women negotiate suggest the persistent strength of gendered divisions of paid labour in shaping geographies of home among communities of mobile professionals (See also Coles & Fechter 2008). For some British women, relocation to Dubai presented a welcome opportunity to focus time and energy on their interpersonal relationships with children, husbands and friends. For others, the reproduction of 'family' weighed heavily on their shoulders alone, resulting in a decline in the equality of a previously intimate marital relationship.

Difficulties in sustaining intimate kin relations over large distances were also in evidence. Technologies, it has been argued, enable transnational social connections to become routine, so that people seem both 'here' and 'there' (Urry 2004). Callon and Law (2004), for instance, argue that text messaging on mobile phones is a way of making oneself present and 'attending to another person' without geographical propinquity. Therefore, presence is not reducible to co-presence, but rather, as Licoppe (2004) suggests, we increasingly have the material resources to support 'absent presence'. Yet, the examples in this chapter and, indeed, some of the stories of friendship from Chapter 5, would suggest that the everyday negotiation of families in transnational spaces is not so easily solved by technology. The evidence from British migrants supports the significance of face-to-face visits, as well as their difficulties, already noted in the transnationalism literature (e.g. Pratt 2012).

Chapter Eight
Our Intimate Lives

The significance of intimacy in our everyday lives is not fully recognised by contemporary geographical scholarship. Geographers have certainly made important critical contributions to interdisciplinary debates on friendship, love, heteronormativity and family (e.g. Bunnell et al. 2012; Harker & Martin 2012; Morrison, Johnston & Longhurst 2012; Oswin & Olund 2010), but, as a discipline, we have tended to view these as separate debates and organise ourselves in sub-disciplinary areas that reinforce this perspective (see also Valentine 2008). It is time, I suggest, to explore more holistically the constitution of intimate subjectivities and the collective cultural discourses about intimacy that shape our everyday lives. There is much more to understand about the way in which intimacy is, as Jamieson (1998: 1) argued so long ago, at the heart of 'meaningful social life'. Since in the UK we now have a popular conceptualisation of an '*intimacy of the self rather than an intimacy of the body*' (Jamieson 1998: 1), intimacy in this book has been deployed not as a synonym for physical closeness, proximity or sexual relationships (though these have been significant in places), but through a broader range of personal relationships among communities, friends and families, as well as couples.

In a book published just prior to my submission of this manuscript, Moss and Donovan (2017) comment on the range of definitions of intimacy that feminist geographers are now employing in their research. In this book, I have explored the relational production, everyday practice and narrative accounts of British migrants' intimate subjectivities, focusing empirically on how Britons resident in the globalising city of Dubai discuss their navigation of a range of interpersonal

Transnational Geographies of the Heart: Intimate Subjectivities in a Globalising City, First Edition. Katie Walsh.
© 2018 John Wiley & Sons Ltd. Published 2018 by John Wiley & Sons Ltd.

relationships. In doing so, I have engaged with broader debates in migration studies about the reconfiguration of intimate subjectivities across and through dialectics of mobility and settlement, here and there, home and away (e.g. Constable 2009; Gorman-Murray 2009; Mai & King 2009; Pratt 2012), and further exemplified the making of privileged 'expatriate' subjectivities in particular postcolonial spaces (e.g. Fechter 2007; Knowles & Harper 2009; Leonard 2010). Increasingly, migration scholars are recognising that migration is shaped through intimacy, and vice versa, with our interpersonal lives informing and responding to *processes* of movement and settlement. Yet, these literatures tend to focus on *either* the negotiation of transnational family life, especially care for elderly parents or children (e.g. Baldassar, Baldock & Wilding 2006; Baldassar & Merla 2014; Constable 2007; Pratt 2012), or the negotiation of sexualities and romance associated with couple relationships (e.g. Constable 2003; Mai & King 2009). While my own ethnographic research on intimacy also suffers from its own incompleteness, emerging as it did without prior planning, I have attempted in this book to demonstrate the potential of trying to understand intimacy more inclusively (see also Moss & Donovan 2017). Removing the automatic association of intimacy as being about sexual and romantic relationships is important since the couple relationship is spatially and discursively privileged, not least in contemporary UK society (Berlant 2000; Jamieson 1998; Wilkinson 2014). A broader conceptualisation of intimacy, then, makes it possible to begin to unpack how our intimate subjectivities are produced through our relations with a range of others in our lives, encompassing friends, families and sexual relationships (including, but not limited to marriage), rather than elevating one kind of relationship as being of paramount concern. Chapter 5, for example, demonstrated that friendships are informed by couple relationships and family ties, including their absence/presence in our lives and their renegotiation in transnational migration contexts. Some single Britons, for instance, enact 'family-like' relations in shared households and develop strong single-sex friendship groups as they navigate Dubai's party scene. Meanwhile, some married migrants quickly establish deeper friendships during a crisis in order to replace family they've left behind in the UK. Arguably, women especially find friendships enormously important in this city to provide emotional and practical support in their parenting routines when husbands are travelling or working long hours.

Arguably, paying attention to migration helps us to observe these kinds of global–local *textures* of intimacy in everyday life, whereby different kinds of relationships with a range of others inform the reproduction of our intimate subjectivities. Movement from one country to another certainly led to a reflexive understanding of the significance of intimacy among my own interlocutors. In other words, I did not have to deliberately set out to investigate intimacy, for the effort of finding or maintaining intimacy, as well as adaptations to (in)formal regulations of intimacy, made it an everyday topic of conversation among British migrants as they sought to navigate the specific topographies of migration in the city they encountered. Since intimacy is culturally coded (Povinelli 2006), Britons carry with them norms and habits as embodied knowledge of how intimacy operates. These then shift as they encounter the informal and formal codes of their new place of residence. Thereby, the observation of migration scholars that mobility and intimacy are co-constituted in everyday migration processes and practices certainly rings true (e.g. Mai & King 2009). As I argued in Chapter 2, however, there are different ways we might acknowledge and understand the *spatialisation* of intimacy, beyond its relation to mobility, and several of these informed my analysis. The next section revisits the framework that I outlined in Chapter 2 for how, and why, we might foreground the spatialisation of intimacy in thinking critically about its significance in everyday life.

Exploring the *Spatialisation* of Intimacy

Intimacy can be understood as a regulatory construct emerging from the co-production of private and public life (Oswin & Olund 2010) and, therefore, intimate relationships are geographically contingent, emerging in place. In Dubai, as I have demonstrated, intimacy is highly regulated through both legislative and spatial means. The UAE's 'decency laws', for example, encode a more extreme notion of heteronormativity than Britons are used to from life in the UK, most notably in the prohibition of sex outside marriage, yet this does not appear to have significant impact, at least collectively, on single migrants' performances of heterosexuality. Knowledge about these regulations, as well as access to routine and emergency contraception, and even the resources to return home if necessary due to pregnancy outside marriage, allow British migrants to circumvent the experience of imprisonment and statelessness that Mahdavi (2016) describes as more widespread in the case of domestic workers. Similarly, the *Kafala* sponsorship system, which strongly regulates low-income migrants' settlement practices by insisting on a certain level of income to bring accompanying family members, as described by Gardner (2011), and is seen in the 'bachelor' communities of low-income south Asians that Elsheshtawy (2010) describes, did not affect British migrants' family lives in the same way. For Britons, their nationality and their (mostly) graduate status was rewarded with a more-than-sufficient salary to ensure partners and children could accompany them to Dubai, as well as financial packages to make it easier for them to do so, including additional allowances for housing, schooling, flights and healthcare. Nevertheless, the *Kafala* sponsorship system also insists on the 'temporariness' of migration to the Gulf, shaping the temporalities of everyday life (e.g. Mohammed & Sidaway 2016) and, arguably, as a consequence, a sense of 'transience' dominated the texture of intimacy in Dubai. Certainly, the production of intimate subjectivities among British migrants was shaped not only by discourses of displacement, but also impermanence, as evident from my discussion of both friendships (Chapter 5) and romance (Chapter 6).

The spatial regulation of intimacy is further enabled and evidenced in Dubai through the privatisation of malls, hotels and other leisure spaces. Club memberships and VCOs further entrench the nationalisation and racialisation of networks that inhabit these privatised leisure spaces. The production of classed and racialised 'expatriate' subjectivities is evident not only in Britons' cross-cultural sexual relations (and their lack of them), but also by their choices as to who to socialise among. Geographical research has consistently demonstrated the necessity of a politicised understanding of intimacy (e.g. Oswin & Olund 2010; Pain & Staeheli 2014) and, in this book, doing so requires a recognition of the postcoloniality of Dubai. It is possible to trace the ongoing articulation of colonial categories of difference being reproduced in British migrants 'expatriate' subjectivities, but also the reworking of these ontologies in relation to Emirati social status and the instabilities of whiteness.

Migrants' intimate subjectivities, reproduced in the localised spaces of the globalising city, are embodied not only in terms of the social locations of the migrant with regard to the intersections of nationality, class, race, sexuality and gender, but also through practice. Empirical material in the chapters provided evidence, for example, of practices of friendship, including helping neighbours in daily parenting or in moments of crisis, sharing perspectives and feelings in coffee morning chats, watching television together in a shared house, or going out together drinking and dancing. David Morgan's (2013) understanding of 'family practices' as the doing of family life, also evidenced in Chapter 7, might therefore be usefully extended to the doing of friendship too. As Carol Smart (2007: 187) suggests, an emphasis on *personal life* as a site of 'connectedness, relationship, reciprocal emotion' can lead us to explore 'all kinds of sociality', rather than focusing on family 'as an inevitable point of reference'. While couple ties are privileged in everyday discourses of intimacy, it is important that critical studies of intimacy recognise the significance of other relationships in people's intimate lives (see Berlant 2000; Giddens 1992; Jamieson 1998). As such, practices of intimacy should not be limited to either those associated with sex or sexuality, nor the emotional disclosure of partners depicted in notions of the 'pure relationship' (Giddens 2002), but, rather, might include a range of ways of 'being close' to another person (see also Jamieson 1998). This is not, either, to privilege geographical proximity in the production of intimacy (Oswin & Olund 2010; Thien 2005).

Finally, the recognition of intimacy as embodied, through the notion of intimate subjectivities, also brings attention to the emotional *geographies* of intimacy. Chapter 4, for instance, demonstrated a range of emotional responses to migration and cross-cultural interaction and encounter (or lack of it) that helped shape British migrants' residence, lifestyles and practices in Dubai, including their choices of joining 'expatriate' organisations and spaces of leisure regulated by membership norms, as seen in Chapter 5. In addition, Chapters 5, 6 and 7 revealed some of the challenges associated with both conducting relationships in a city so strongly associated with mobility, and with sustaining relationships with family and friends back in the UK, and these were narratives laden with emotion, especially the women's accounts. While existing work on migrants' sexualities and gendered identities has also provided insight into the emotional geographies of migration arising from transformations in subjectivity (e.g. Gorman-Murray 2009), attention to intimacy can further contribute to this project. This involves going beyond 'love', which has been, arguably, a privileged emotion in analyses of intimacy across disciplines, to explore instead the myriad of emotions through which intimate subjectivities are enacted. Alongside love, other emotions that can be traced in the narratives of British migrants included in this book are desire, fear, hope, disgust, affection, disappointment, resentment, and loss, among others.

In this way, then, I have demonstrated that a critical analysis of the *geographies* of intimacy might productively attend to the ways in which intimate subjectivities are embodied, emplaced and co-produced across binaries of public/private and local/global space.

Changing Gulf Subjectivities: *Temporalities* of Intimacy

The production of intimate subjectivities is not only spatially contingent, but also dependent on temporal particularities. As I described in Chapter 3, this book draws on an ethnographic study of British migrant belonging undertaken between November 2002 and April 2004, a period during which I lived in Dubai to complete my doctoral fieldwork. As such, although I continued to visit Dubai and the wider GCC region in the subsequent decade to engage with comparative postdoctoral research (see Walsh 2014), and also to lead undergraduate field classes, this book emerges from the earlier sustained period of fieldwork. This clearly matters because it locates my analysis of British migrants' intimate subjectivities not only in place, but also in time. While in this book, then, I have foregrounded the spatialisation of intimacy, the analysis has also been specific about the temporalities of the fieldwork site. Dubai was certainly increasing in global significance at this time, but was not yet the Dubai of today. The term global*ising* city (see, for example, Yeoh 1999) was certainly pertinent. A sub-sample of interviewees had lived in Dubai for twenty or more years and had witnessed not only huge changes in the fabric of the city, but also a large and rapid population increase of migrants, including Britons. They remembered a Dubai in which they had been pioneering families on 'hardship postings' among a small, distinct and valued community of experts assisting in the transformation of the Emirates often through vital infrastructural projects. Other interviewees were part of the more recent wave of migrants, may have travelled alone, and were often attracted by the consumption culture they could now engage in that would have been unthinkable just a decade previously. The research therefore captures a particular moment in the globalisation of Dubai, when the city was being re-imagined through an 'opening up' to global flows of investment, people and things. This is not to forget, however, the much longer histories of travel and settlement across the region, as well as the twentieth-century transnationalism of the Arab Gulf that already marked the UAE as thoroughly diverse (see Chapter 1).

A few examples of concrete changes in the urban fabric and, consequently, 'expatriate' routines may be useful in order to explore the temporality of my field-work further. Firstly, the geography of British migrants' housing has changed in Dubai. Davidson (2008) discusses how residential developments were key to the economic diversification strategies of Dubai at this time such that, by 2006, over 30,000 new homes had been built in residential free zones, mostly bought by expatriates, and often sold out within days. As a result of this success, Davidson suggests, Jumeirah Beach Residence in Dubai Marina was later launched, alongside smaller developments including The Greens Community and Arabian Ranches, as well as world-famous mega-projects such as Palm Island. This increased demand such that further developments were designed, including the Burj Dubai and Business Bay project, as well as cheaper developments aimed at the large middle-class South Asian community, such as International City and the Lagoons. The British migrants I spoke with until I left in 2004 were largely hesitant about property ownership. Indeed, I went on a tour of The Greens Community with the Dubai Adventure Mums and heard many cautionary tales of the uncertainties of legislation and cheap building materials. Yet, when I returned in 2006, many had started to buy villas and apartments in these new neighbourhoods. As a result, British migrants were no longer confined to Jumeirah, Umm Suqeim, Sheikh Zayed Road, and select other apartment blocks in Bur Dubai and on the eastern side of the creek. Rather, they were distributed over a much larger city, in larger numbers, and consequently using new schools, malls, supermarkets and leisure spaces during their residence. For instance, the growth of the city's affluent population enabled 'super-clubs' to open since 2004 and attract international DJs and serious clubbers from all over the Middle East (e.g. three-floor club Trilogy in the Madinat Jumeirah complex, the Peppermint Lounge Club in the Fairmont Hotel or MIX in the Grand Hyatt).

Another example is that of 'The Dubai Adventure Mums', whose activities I detailed in Chapter 5. This group no longer exists in Dubai and mums' groups have transformed since the time of my research, reflecting the growth and development of the city. The website 'expatmum.com' also began in 2001, but as a company that the owner franchised as a global brand. During my fieldwork it was already becoming a major community organisation with larger events: monthly coffee mornings (attracting 50–70 women); 'fashion and film' events at the Mercato shopping mall (attracting 80–120 women); and a Ladies Fashion evening at the Fairmont Hotel. Drury became a notable expert on expatriate family life in Dubai, authoring the *Family Explorer Guide* (Drury 2004), writing a monthly feature for

Living Magazine, and often contributing or being quoted in other expatriate newspapers and magazines. Today the website expatwoman.com, which was just emerging in 2003–2004, coordinates multiple groups of mums across the city. This development can be traced to the changing nature of neighbourhoods described above and sheer increase in numbers of expatriate mums living in the city: it became necessary for multiple groups to be established and conveniently located in different residential areas, and this was accompanied by an increased provision of cafes and soft-play facilities. Now there are ExpatBumps and Babes or ExpatMums and Tots groups taking place in the Marina area and Ibn Battuta Mall on several days each week. Nevertheless, many of the reasons for the existence of mums' groups remain the same, as evident from the website of ExpatMums (www.expatwoman. com/dubai/monthly_events_whats_on.aspx#CoffeeMorning). On the site, social groups for mothers are introduced with welcoming photographs of their 'hosts' and statements resonate strongly with the discussion of gendered intimate subjectivities that emerged in Chapter 5: 'We completely understand how it can feel quite isolating in Dubai at times, and so if you're feeling that way it's time to come and meet this friendly group of ladies who are all in the same situation as you.' Or, 'In Dubai, where so many of us are without family support, having a baby can be a daunting prospect, and finding a wonderful group of friends can make a huge difference. It's nice and friendly, so please don't be shy!' The purpose of these groups therefore echoes that of the groups in existence a decade previously, suggesting some continuity in the experiences of motherhood and production of 'expatriate' femininities in Dubai. Nevertheless, the extent of this urban change raises new questions about intimacy.

Especially salient are questions about cross-cultural intimacies. What does it mean for the 'boundaries' of expatriate lives (Fechter 2007) when privileged migrants find their status as white expatriates in the global city diminishing? Does the shared experience of motherhood allow for friendships between middle-class Filipino women and British women raising children in mixed neighbourhoods such as Jumeirah Beach Residence? Does using the metro side-by-side enable spontaneous exchanges between migrants of different income levels and nationalities or shared moments when the social hierarchy breaks down completely? However temporary or hesitant such equalities might seem at first, they are signs of hope that the racial and class stratification that shaped intimate subjectivities during my fieldwork period may be thoroughly reworked in the future.

Another major implication of conducting the fieldwork during the early 2000s is that although email, text messaging and international phone calls were available in Dubai, the kinds of instantaneous visual communications technologies that are now taken for granted – Facebook, Skype, Instagram, Twitter and What's App – were unavailable or not yet widely adopted. In this way the transnational social landscape was rather different than it must be for today's Britons who are supported by Wi-Fi at home and in every shopping mall, hotel or other leisure space. This means that this book is not a study of the use of communication technologies, which might otherwise be expected in an analysis of transnational migrants' lives and a globalising world. Nonetheless, I have found this temporal and technological demarcation helpful, since otherwise the temptation would have been to focus much more on the practices of transnationalism that such technologies enable or prevent, and to ignore the emotional negotiation of relationships during emplacement. Arguably, this would have helped to further reproduce the idea of skilled migrants as detached from place. Instead, a focus on the everyday negotiation of relationships, primarily those within Dubai, highlights the feelings, inequalities and embodied performances of our intimate subjectivities. Still, the massive explosion in the use of ICT by migrants to sustain a sense of co-presence among their transnational family members does need further attention. Recently, Baldassar et al. (2016: 133) argued that while migration has always challenged the premise that strong relationships require physically proximate interactions, a 'recent, startling emergence of a new social environment of ubiquitous connectivity' is transforming caring relations and practices in transnational families.

Intimacy, Belonging and Home: Rethinking the Global City

In the days prior to my departure from Dubai, a countdown in lipstick on my bedroom mirror marked the days I had left to spend time with my friends there. They organised a barbecue on the night before I flew home and my closest friend presented me with a photo album recording the previous months. I laughed through tears of sadness as I turned the pages of the album in the company of the people I had been sharing my life with over the previous year. Throughout the flight 'home', I tried not to think, but films and meals failed to distract me. Emails, text messages, photos, memories and visits exerted an emotional pull to Dubai for months, perhaps years, afterwards. It was these experiences that really convinced me of the significance of intimacy in shaping transnational geographies of the heart and migrant subjectivities. As Karen Till (2001: 46) reflects, 'returning home' after fieldwork can be disorientating: 'I felt almost schizophrenic, torn between worlds, cultures, sets of social relations and selves.' My feeling of belonging in Dubai was amplified by a sense of non-belonging in the academic world I now encountered in libraries, conferences, seminars rooms, as well as the loneliness of the 'writing up' stage. Academics asked me how I coped with living 'there', – 'It's all those horrible hotels and shopping isn't it' – or questioned my 'hanging out with such airheads', while I was struggling to cope with living back 'home' and leaving behind my new friendships. I share these observations not to claim ethnographic authority for this book, but to reassert the centrality of intimacy in shaping migration and everyday life.

The links between the production of 'family' and the domestic home space are well established (e.g. Bowlby, Gregory & McKie 1993; Christensen, James &

Jenks 2000; Dowling & Power 2012; Harden et al. 2013; Morgan 2013). However, the way in which intimate subjectivities play a significant role in broader geographies of the heart – including those narrated around home and belonging – has not been fully developed or articulated. Existing calls have urged us to consider love and sexuality as part of an emotional turn in migration studies (e.g. Mai & King 2009) and we know that mobility can lead to the production of new subjectivities whereby our queered or gendered selves can be more fully expressed in the safety of different citizenship norms or in the freedom that can come from being away from family and kin (e.g. Gorman-Murray 2009). Arguably, too, notions of diaspora, transnational social fields, and transnational families also have in common an understanding of their constitution through transnational practices of relationships over time and space (e.g. Pratt 2012). Empirical research has also extensively mapped the reconfiguration of intimate subjectivities in situations of global work and cross-cultural marriage (e.g. Constable 2009), not least in the GCC region (Gardner 2011; Mahdavi 2016). Yet, the way in which intimacy impacts upon every aspect of our daily lives has been obscured, I believe, by the focus on individual types of interpersonal relationship. When we begin to map a range of interpersonal relationships that contribute to the making of our intimate subjectivities, we begin to see how the textures of intimacy in a particular place might shape both how and where people feel at home (or not) in this world. As a result, it is my belief that any attempt to explore the spatialisation of everyday life, whether for transnational migrants or not, needs to encompass an understanding of intimacy in the production of our selves. Our intimate lives hold insights into the way in which our subjectivities are folded into space, not only through their governance, but also through the embodiment and emplacement of interpersonal relationships in the everyday practices of intimacy through which home and belonging emerge.

Rethinking global cities research from a postcolonial perspective we are asked to 'creatively reconstruct the ways in which globalization emerges in cities in all sorts of ways' (Collins 2016: 289). Certainly, the 'globalness' of the GCC region might be recognised more clearly if we were to acknowledge the transnational connections of its millions of temporary migrant workers (Price & Benton-Short 2007). Indeed, Ananya Roy (2009: 828, drawing on Mitchell, Chakrabarty & Robinson) comments on Dubai as a specific illustration of how we might take seriously 'the emergence of the modern outside the geography of the West', so that 'the standard geographies of core and periphery are disrupted and dislocated'. She explains: 'In such a world, Dubai is the lodestone of desires and aspirations, the icon of supermodernity in the backbreaking trudge of transnational migration from the villages of Egypt, Bangladesh, Indonesia and Pakistan.'

One of the implications of Roy's (2009) insistence on the 'worldliness' of multiple urban sites such as Dubai is her argument that although the modernities of these cities are distinctive, they can also inform our analysis of other places. Supporting this, the translocalities through which British migrants' intimate subjectivities are constructed became evident in the empirical chapters of this book. Their accounts of community, romance, marriage and friendship tell us not only about the (in)formal regulation of intimate life in Dubai, but also about intimacy norms in the UK. The ethnographic evidence, then, asserts the relational spatialities of intimacy in the global city. Yet, there is much more analytical work to be done in exploring the spatialisation of intimacy, not least the moments when transnational cultural flows of ideas, objects, media, emotions and technologies inform, or are contested by, our embodied enactment of intimate subjectivities. How do intimacy norms, practices, discourses and mobilities play out in the many varied interactions of the global city? As Roy (2009: 829) continues:

> It is known how to map the 'global' through Darwinian hierarchies of city-regions; much less is known about the complex connections, exchanges, and references through which cities (everywhere) are worlded. The world is not flat, and it is time to produce a more contoured knowledge of its cities.

The emergence of a body of deep ethnographic work on Gulf subjectivities over the last decade is an exciting part of this development (e.g. Elsheshtawy 2010; Mahdavi 2016).

The title of this book – 'transnational geographies of the heart' – is a plea to geographers to take intimate subjectivities more seriously in conceptualising the spatialities of globalisation and migration. Focusing on our intimate lives does not involve a turn away from global processes. Rather, geographers have emphasised the need to think about the spatialisation of intimacy without recourse to scalar understandings of space (e.g. Oswin & Olund 2010; Pain & Staeheli 2014; Pratt & Rosner 2012). It is not, then, about viewing British migrants' intimate subjectivities as both local and global, since localised and globalised relations cannot be distinguished as separate (see Amin 2002). It is not, either, about an alternate analytic to the economic framework that has dominated globalisation theory, since intimacy is entangled with global work (e.g. Pratt 2012) and dichotomies of public/private are thoroughly untenable (see Oswin & Olund 2010). Instead, attention to migrants' intimate subjectivities requires us to listen to the complexity of how they are co-produced with globalisation and all its inequalities. As we have seen, the privilege of the British 'expatriate' migrant status contrasts starkly with the status conferred on migrants of many other nationalities in Dubai, but also highlights a further paradox in the politics of current British immigration debates. Intimate subjectivities are, therefore, significant in understanding not only people's experiences of the everyday in global(ising) cities, but to an analysis of contemporary articulations of global urbanity.

References

Ahmed, Sara (2004). *The Cultural Politics of Emotion*. Edinburgh: Edinburgh University Press.

Aitken, Stuart (2005). The awkward spaces of fathering. In Bettina van Hoven & Kathrin Hörschelmann (eds), *Spaces of Masculinities*. London: Routledge, pp. 222–236.

Akass, Kim & McCabe, Janet (eds) (2004). *Reading Sex and the City*. London: I B Tauris and Co. Ltd.

Al Abed, Ibrahim & Hellyer, Peter (eds) (2001). *United Arab Emirates: A New Perspective*. London: Trident Press Ltd.

Al-Fahim, Mohammed (1995). *From Rages to Riches: A Story of Abu Dhabi*. London: The London Centre of Arab Studies.

Ali, Syed (2010). *Dubai: Gilded Cage*. London: Yale University Press.

Ali, Syed (2011). Going and coming and going again: Second-generation migrants in Dubai. *Mobilities*, 6(4): 553–568. DOI: 10.1080/17450101.2011.603947.

Allan, Graham (1989). *Friendship: Developing a Sociological Perspective*. London: Harvester Wheatsheaf.

Allan, Graham (1996). *Kinship and Friendship in Modern Britain*. Oxford: Oxford University Press.

Amin, Ash (2002). Spatialities of globalisation. *Environment and Planning A*, 34: 385–399. DOI: 10.1068/a3439

Anthias, Floya (2012). Hierarchies of social location, class and intersectionality: Towards a translocational frame. *International Sociology*, 28(1): 121–138. DOI: 10.1177/0268580912463155

Antonsich, Marco (2010). Searching for belonging – An analytical framework. *Geography Compass*, 4(6): 644–659. DOI: 10.1111/j.1749-8198.2009.00317.x

Armstrong, Gary (1993). Like that Desmond Morris? In Dick Hobbs & Tim May (eds), *Interpreting the Field: Accounts of Ethnography*. Oxford: Oxford University Press, pp. 3–43.

Transnational Geographies of the Heart: Intimate Subjectivities in a Globalising City, First Edition. Katie Walsh.
© 2018 John Wiley & Sons Ltd. Published 2018 by John Wiley & Sons Ltd.

Bagaeen, Samer (2007). Brand Dubai: the instant city; or the instantly recognizable city. *International Planning Studies*, 12(2): 173–197. DOI: 10.1080/13563470701486372

Baldassar, Loretta (2008). Missing kin and longing to be together: Emotions and the construction of co-presence in transnational relationships. *Journal of Intercultural Studies*, 29(3): 247–266. DOI: 10.1080/07256860802169196.

Baldassar, Loretta, Baldock, Cora & Wilding, Raelene (2006). *Families Caring Across Borders: Migration, Ageing and Transnational Caregiving*. Basingstoke: Palgrave MacMillan.

Baldassar, Loretta & Merla, Laura (eds) (2014). *Transnational Families, Migration and the Circulation of Care: Understanding Mobility and Absence in Family Life*. London: Routledge.

Baldassar, Loretta, Nedelcu, Mihaela, Merla, Laura & Wilding, Raelene (2016). IT-based co-presence in transnational families and communities: Challenging the premise of face-to-face proximity in sustaining relationships. *Global Networks*, 16: 133–144. DOI: 10.1111/glob.12108

Batnitzky, Adina, McDowell, Linda & Dyer, Sarah (2008). A middle-class global mobility? The working lives of Indian men in a west London hotel. *Global Networks*, 8(1): 51–70.

Bauman, Zygmunt (2003). *Liquid Love*. Cambridge: Polity Press.

Beaverstock, Jonathan (1996). Migration, knowledge and social interaction: Expatriate labour within investment banks. *Area*, 28: 459–470.

Beaverstock, Jonathan (2002). Transnational elites in global cities: British expatriates in Singapore's financial district. *Geoforum*, 33. 525–538. http://dx.doi.org/10.1016/S0016-7185(02)00036-2

Beaverstock, Jonathan (2005). Transnational elites in the city: British highly-skilled inter-company transferees in New York City's financial district. *Journal of Ethnic and Migration Studies*, 31: 245–268. DOI: http://dx.doi.org/10.1080/1369183042000339918

Beaverstock, Jonathan (2011). Servicing British expatriate 'talent' in Singapore: Exploring ordinary transnationalism and the role of the 'expatriate' club. *Journal of Ethnic and Migration Studies*, 37(5): 709–728. DOI: 10.1080/1369183X.2011.559714

Beaverstock, Jonathan & Bordwell, James (2000). Negotiating globalization, transnational corporations and global city financial centres in transient migration studies. *Applied Geography*, 20: 277–304. http://dx.doi.org/10.1016/S0143-6228(00)00009-6

Beck, Ulrich (1995). *The Normal Chaos of Love*. Cambridge: Polity Press.

Beck, Ulrich & Beck-Gernsheim, Elisabeth (2014). *Distant Love: Personal Life in the Global Age*. Cambridge: Polity Press.

Beck, Ulrich, Giddens, Antony & Lash, Scott (1994). *Reflexive Modernization: Politics, Tradition and Aesthetics in the Modern Social Order*. Stanford University Press: California.

Behdad, Ali (1994). *Belated Travellers: Orientalism in the Age of Colonial Dissolution*. Durham NC: Duke University Press.

Bell, David & Valentine, Gill (eds) (1995). *Mapping Desire: Geographies of Sexualities*. London: Routledge.

Bell, Diane (1993). 'Yes Virginia, there is a feminist ethnography.' In Diane Bell, Pat Caplan & Wazir Jahan Karim (eds), *Gendered Fields: Women, Men and Ethnography*, pp. 28–43. London: Routledge.

Bell, Diane, Caplan, Pat & Wazir Jahan Karim (eds) (1993). *Gendered Fields: Women, Men and Ethnography*. London: Routledge.

Bell, Sandra & Coleman, Simon (1999). The anthropology of friendship: Enduring themes and future possibilities. In Sandra Bell & Simon Coleman (eds), *The Anthropology of Friendship*. Oxford: Berg, pp. 1–19.

Benson, Michaela (2010). The context and trajectory of lifestyle migration: The case of the British residents of southwest France. *European Societies*, 12(1): 45–64. DOI: 10.1080/14616690802592605

Benson, Michaela & O'Reilly, Karen (2009). Migration and the search for a better way of life: A critical exploration of lifestyle migration. *The Sociological Review*, 57: 608–625. DOI: 10.1111/j.1467-954X.2009.01864.x

Benton-Short, Lisa, Price, Marie D. & Friedman, Samantha (2005). Globalization from below: The ranking of global immigrant cities. *International Journal of Urban and Regional Research*, 29(4): 945–959. DOI: 10.1111/j.1468-2427.2005.00630.x

Berg, Lawrence & Longhurst, Robyn (2003). Placing masculinities and geography. *Gender, Place and Culture*, 10(4): 351–360. http://dx.doi.org/10.1080/0966369032000153322

Berlant, Lauren (1998). Intimacy: A Special Issue. *Critical Inquiry*, 24(2): 281–288. DOI: 10.1086/448875.

Berlant, Lauren (ed.) (2000). *Intimacy*. Chicago: University of Chicago Press.

Bernard, H. Russell (1994). *Research Methods in Anthropology* (2nd edition). London: Sage.

Binnie, J. (2004). *The Globalization of Sexuality*. London: Sage.

Blunt, Alison (1999). Imperial geographies of home: British women in India, 1886–1925. *Transactions of the Institute of British Geographers*, NS 24: 421–440. DOI: 10.1111/j.0020-2754.1999.00421.x

Blunt, Alison (2008). Cultural geographies of migration: Mobility, transnationality and diaspora. *Progress in Human Geography*, 31: 684–694. DOI: 10.1177/0309132507078945

Blunt, Alison & Robyn Dowling (2006). *Home*. London: Routledge.

Boccagni, Paolo & Baldassar, Loretta (2015). Emotions on the move: Mapping the emergent field of emotion and migration. *Emotion, Space and Society*, 16: 73–80. http://dx.doi.org/10.1016/j.emospa.2015.06.009

Bondi, Liz (2005). Making connections and thinking through emotions: Between geography and psychotherapy. *Transactions of the Institute of British Geographers*, 30: 433–448.

Bonnett, Alistair (2000). *White Identities: Historical and International Perspectives*. Harlow: Prentice Hall.

Bowlby, Sophie (2011). Friendship, co-presence and care: Neglected spaces. *Social and Cultural Geography*, 12, 605–622. DOI: 10.1080/14649365.2011.601264

Bowlby, Sophie, Gregory, Susan & McKie, Linda (1993). 'Doing' home: Patriarchy, caring and space. *Women's Studies International Forum*, 20(3): 343–350. http://dx.doi.org/10.1016/S0277-5395(97)00018-6

Brinkworth, Lisa (2001a). The Dubai Dolls. *Cosmopolitan, November* 2001, pp. 54–58.

Brinkworth, Lisa (2001b). The girls who party in a fool's paradise. *The Daily Mail*, October 15 2001, pp. 20–21.

Browne, Kath, Lim, Jason & Brown, Gavin (2007). *Geographies of Sexualities: Theory, Practices and Politics*. Aldershot: Ashgate.

Bryceson, Deborah & Vuorela, Ulla (2002). Transnational families in the twenty-first century. In Debroah Bryceson & Ulla Vuorela (eds), *The Transnational Family: New European Frontiers and Global Networks*. New York: Berg, pp. 3–30.

Buckley, Michelle (2012). From Kerala to Dubai and back again: Migrant construction workers and the global economic crisis. *Geoforum*, 43(2): 250–259. http://dx.doi.org/10.1016/j.geoforum.2011.09.001

Bunnell, Tim, Yea, Sallie, Peake, Linda, Skelton, Tracey & Smith, Monica (2012). Geographies of friendships. *Progress in Human Geography*, 36(4): 490–507. DOI: 10.1177/0309132511426606

Burawoy, Michael, Blum, Joseph A., George, Sheba, Gille, Zsuzsa & Thayer, Millie (2000). *Global Ethnography: Forces, Connections, and Imaginations in a Postmodern World*. California: University of California Press.

Butler, Ruth & Parr, Hester (1999). *Mind and Body Spaces: Geographies of Illness, Impairment and Disability*. London: Routledge.

Butz, David & Besio, Kathryn (2004). The value of autoethnography for field research in transcultural settings. *The Professional Geographer*, 56: 350–360. DOI: 10.1111/j.0033-0124.2004.05603004.x

Callard, Felicity (1998). The body in theory. *Environment and Planning D: Society and Space*, 16: 387–400. DOI: 10.1068/d160387

Callon, Michel & Law, John (2004). Introduction: Absence – presence, circulation, and encountering in complex space. *Environment and Planning D: Society and Space*, 22(1): 3–11. DOI: 10.1068/d313

Canfield, Jack, Hansen, Mark, Shimoff, Marci & Hawthorne, Jennifer (eds) (1996). *Chicken Soup for the Woman's Soul*. Florida: Health Communications Ltd.

Chapman, Tony (2001). There's no place Like home. *Theory, Culture and Society*, 18(6): 135–146. DOI: 10.1177/02632760122052084

Christensen, Pia, James, Allison & Jenks, Chris (2000). Home and movement: Children constructing 'family time'. In Sarah Holloway & Gill Valentine (eds), *Children's Geographies: Playing, Living and Learning*. London: Routledge, pp. 139–155.

Coles, Anne & Fechter, Anne-Meike (eds) (2008). *Beyond the Incorporated Wife: Gender Relations among Mobile Professionals*. London: Routledge.

Coles, Anne & Jackson, Peter (2007). *Windtower*. London: Stacey International.

Coles, Anne & Walsh, Katie (2010). From 'Trucial States' to 'Postcolonial City'? The imaginative geographies of British expatriates in Dubai. *Journal of Ethnic and Migration Studies* (Special Issue), 36: 1317–1334. DOI: 10.1080/13691831003687733

Collins, Francis L. (2016). What if the global city was postcolonial? *Dialogues in Human Geography*, 6(3): 287–290. http://dx.doi.org/10.1177/2043820616676566

Conradson, David & Latham, Alan (2005a). Transnational urbanism: Attending to everyday practices and mobilities. *Journal of Ethnic and Migration Studies*, 31: 227–233. DOI: 10.1080/1369183042000339891

Conradson, David & Latham, Alan (2005b). Friendship, networks and transnationality in a world city: Antipodean transmigrants in London. *Journal of Ethnic and Migration Studies*, 31: 287–305. DOI: 10.1080/1369183042000339936

Conradson, David & Latham, Alan (2007). The affective possibilities of London: Antipodean transnationals and the overseas experience. *Mobilities*, 2(2): 231–254. DOI: 10.1080/17450100701381573

Conradson, David & McKay, Deirdre (2007). Translocal subjectivities: Mobility, connection, emotion. *Mobilities*, 2(2): 167–174. DOI: 10.1080/17450100701381524

Constable, Nicole (2003). *Romance on a Global Stage: Pen Pals, Virtual Ethnography, and 'Mail Order' Marriages*. Berkeley: University of California Press.

Constable, Nicole (2007). *Maid to Order in Hong Kong: Stories of Migrant Workers*. Ithaca, New York: Cornell University Press.

Constable, Nicole (2008). Introduction: Distant divides and intimate connections. *Critical Asian Studies*, 40(4): 551–566. DOI: 10.1080/14672710802505299.

Constable, Nicole (2009). The commodification of intimacy: Marriage, sex and reproductive labor. *Annual Review of Anthropology*, 38: 49–64.

Constable, Nicole (2014). *Born Out of Place: Migrant Mothers and the Politics of International Labor*. Berkeley: University of California Press.

Conway, Daniel & Leonard, Pauline (2014). *Migration, Space and Transnational Identities: The British in South Africa*. Basingstoke: Palgrave MacMillan.

Corber, Robert & Valocchi, Stephen (eds) (2006). *Queer Studies: An Interdisciplinary Reader*. Oxford: Blackwell.

Craik, Jennifer (1997). The culture of tourism. In Christopher Rojek & John Urry (eds), *Touring Cultures: Transformations of Travel and Theory*. London: Routledge, pp. 113–136.

Cranston, Sophie (2016). Producing migrant encounter: Learning to be a British expatriate in Singapore through the global mobility industry. *Environment and Planning D: Society and Space*, 34(4): 655–671. DOI: 10.1177/0263775816630311

Crocetti, Gina (2000). *Culture Shock! United Arab Emirates: A Guide to Customs and Etiquette*. London: Kuperard.

Crouch, David (ed.) (1999). *Leisure/Tourism Geographies: Practices and Geographical Knowledge*. London: Routledge.

Croucher, Martin (2011). Dubai Country Club 'unlikely to reopen'. Deposits have been returned to prospective members of the beloved institution. *The National*, 25 December 2001: https://www.thenational.ae/uae/dubai-country-club-unlikely-to-reopen-1.378935

Davidson, Christopher (2008). *Dubai: The Vulnerability of Success*. New York: Columbia University Press.

Davis, Mike (2005). Metropolitan disorders 1: Fear and money in Dubai. *New Left Review*, 41: 47–68.

De Lyser, Dydia & Starrs, Paul (eds) (2001). Doing fieldwork. *Geographical Review* (Special Issue), 91.

Denscombe, Martyn (1998). *The Good Research Guide*. Buckingham: Open University Press.

Dowling, Robyn & Power, Emma (2012). Sizing home, doing family in Sydney, Australia. *Housing Studies*, 27(5): 605–619. http://dx.doi.org/10.1080/02673037.2012.697552

Drury, Jane (2004). *Family Explorer Guide Abu Dhabi Dubai* (2nd edition). Dubai: Explorer Publishing Ltd.

Duncan, James & Gregory, Duncan (eds) (1998). *Writes of Passage: Reading Travel Writing*. London: Routledge.

Dunn, Kevin (2009). Embodied transnationalism: Bodies in transnational spaces. *Population, Space and Place*, 16(1): 1–9.

Edensor, Tim (2001). *National Identity, Popular Culture and Everyday Life*. Oxford: Berg.

Ehrenreich, Barbara & Hochschild, Arlie (eds) (2003). *Global Woman: Nannies, Maids, and Sex Workers in the New Economy*. London: Granta

Elsheshtawy, Yasser (2008a). Navigating the spectacle: Landscapes of consumption in Dubai. *Architectural Theory Review*, 13(2): 164–187.

Elsheshtawy, Yasser (2008b). Transitory sites: Mapping Dubai's 'forgotten' urban spaces. *International Journal of Urban and Regional Research*, 32(4): 968–988.

Elsheshtawy, Yasser (2010). *Dubai: Behind an Urban Spectacle*. London: Routledge.

England, Kim (1994). Getting personal: Reflexivity, positionality, and feminist research. *Professional Geographer*, 46: 80–89. http://dx.doi.org/10.1111/j.0033-0124.1994.00080.x

Eve, Michael (2002). Is friendship a sociological topic? *European Journal of Sociology*, 43: 386–409.

Explorer Publishing (2002). *Dubai Zappy Explorer*. Dubai: Explorer Publishing.

Fabian, Johannes (1983). *Time and the Other: How Anthropology Makes Its Object*. New York: Columbia University Press.

Fargues, Philippe (2011). Immigration without inclusion: Non-nationals in nation-building in the Gulf states. *Asian and Pacific Migration Journal*, 20(3–4): 273–292.

Farrer, James (2010). A foreign adventurer's paradise? Interracial sexuality and alien sexual capital in reform-era Shanghai. *Sexualities*, 13: 69–95.

Farrer, James (2011). Global nightscapes in Shanghai as ethnosexual contact zones. *Journal of Ethnic and Migration Studies*, 37(5): 747–764.

Farrer, James (2012). 'New Shanghailanders' or 'New Shanghainese': Western expatriates' narratives of emplacement in Shanghai. In Anne-Meike Fechter & Katie Walsh (eds), *The New Expatriates: Postcolonial Approaches to Mobile Professionals*. London: Routledge, pp. 23–40.

Fechter, Anne-Meike (2005). The Other stares back: Experiencing whiteness in Jakarta. *Ethnography*, 6(1): 87–103.

Fechter, Anne-Meike (2007). *Transnational Lives: Expatriates in Indonesia*. Hampshire: Ashgate.

Fechter, Anne-Meike (2008). From 'Incorporated Wives' to 'Expat Girls': A new generation of expatriate women? In Anne Coles & Anne-Marie Fechter (eds), *Gender and Family among Transnational Professionals*. London: Routledge, pp. 193–210.

Fechter, Anne-Meike (2010). Gender, empire, global capitalism: Colonial and corporate expatriate wives. *Journal of Ethnic and Migration Studies*, 36(8), 1279–1297.

Fechter, Anne-Meike (2016). Mobility, white bodies and desire: Euro-American women in Jakarta. *The Australian Journal of Anthropology*, 27(1), 66–83.

Fechter, Anne-Meike & Walsh, Katie (2010). Introduction to Special Issue: Examining 'expatriate' continuities: Postcolonial approaches to mobile professionals. *Journal of Ethnic and Migration Studies*, 36: 1197–1210. DOI: 10.1080/13691831003687667

Fechter, Anne-Meike & Walsh, Katie (eds) (2012). *The new expatriates: Postcolonial approaches to mobile professionals*. London: Routledge.

Fehr, Beverley (2000). The life cycle of friendship. In Clyde Hendrick & Susan S. Hendrick (eds), *Close Relationships: A Sourcebook*. http://dx.doi.org/10.4135/9781452220437.n6

Finch, Janet (2007). Displaying families. *Sociology*, 41(1): 65–81. http://dx.doi.org/10.1177/0038038507072284

Finch, Janet & Mason, Jennifer (1993). *Negotiating Family Responsibilities*. London: Routledge.

Finch, Tim, with Holly Andrew & Maria Latorre (2010). *Global Brit: Making the Most of the British Diaspora*. London: Institute of Public Policy Research.

Findlay, Allan M. (1988). From settlers to skilled transients: The changing structure of British international migration. *Geoforum*, 19: 401–410. http://dx.doi.org/10.1016/S0016-7185(88)80012-5

Findlay, Allan M., Li, Lin, Jowett, John & Skeldon, Ronald (1996). Skilled international migration and the global city: A study of expatriates in Hong Kong. *Transactions of the Institute of British Geographers*, 21(1): 49–61. DOI: 10.2307/622923

Fortier, Anne-Marie (2000). *Migrant Belongings: Memory, Space, Identity*. Oxford and New York: Berg.

Fortier, Anne-Marie (2003). Making home: Queer migrations and motions of attachment. In Sara Ahmed, Claudia Castaneda, Anne-Marie Fortier & Mimi Sheller (eds), *Uprootings/Regroundings: Questions of Home and Migration*. Oxford and New York: Berg, pp. 115–135.

Frankenberg, Ruth (1993). *White Women, Race Matters: The Social Construction of Whiteness*. Minneapolis: University of Minnesota Press.

Gabb, Jacqui (2010). *Researching Intimacy in Families*. Basingstoke: Palgrave MacMillan.

Gaetano, Arianne (2008). Sexuality in diasporic space: Rural-to-urban migrant women negotiating gender and marriage in contemporary China. *Gender, Place and Culture*, 15: 629–645. http://dx.doi.org/10.1080/09663690802518545

Gardner, Andrew (2008). Strategic transnationalism: The indian diasporic elite in contemporary Bahrain. *City & Society*, 20(1): 54–78. DOI: 10.1111/j.1548-744X.2008.00005.x

Gardner, Andrew (2010). *City of Strangers*. Ithaca: ILR Press.

Gardner, Andrew (2011). Gulf migration and the family. *Journal of Arabian Studies*, 1(1): 3–25.

Giddens, Anthony (1992). *The Transformation of Intimacy*. Cambridge: Polity Press.

Gorman-Murray, Andrew (2006). Gay and lesbian couples at home: Identity work in domestic space. *Home Cultures*, 3(2): 145–168. http://dx.doi.org/10.2752/174063106778053200

Gorman-Murray, Andrew (2007). Rethinking queer migration through the body. *Social and Cultural Geography*, 8: 105–121. DOI: 10.1080/14649360701251858

Gorman-Murray, Andrew (2008). Masculinity and the home: A critical review and conceptual framework. *Australian Geographer*, 39: 367–379. DOI: 10.1080/00049180802270556

Gorman-Murray, Andrew (2009). Intimate mobilities: Emotional embodiment and queer migration. *Social & Cultural Geography*, 10(4):441–460. DOI:10.1080/14649360902853262

Gray, Breda (2008). Putting emotion and reflexivity to work in researching migration. *Sociology*, 42: 935–952. DOI: 10.1177/0038038508094571.

Haldrup, Michael & Larsen, Jonas (2009). *Tourism, Performance and the Everyday: Consuming the Orient*. London: Routledge.

Hall, Stuart (1996). When was 'the post-colonial'? Thinking at the limit. In Iain Chambers & Lidia Curti (eds), *The Post-colonial Question: Common Skies, Divided Horizons*. London and New York: Routledge, pp. 242–260.

Harden, Jeni, Backett-Milburn, Kathryn, MacLean, Alice, Cunningham-Burley, Sarah & Jamieson, Lynn (2013). Home and away: Constructing family and childhood in the context of working parenthood. *Children's Geographies*, 11(3): 298–310. http://dx.doi.org/10.1080/14733285.2013.812274

Hardill, Irene (1998). Gender perspectives on British expatriate work. *Geoforum*, 29: 257–268. http://dx.doi.org/10.1016/S0016-7185(98)00012-8

Hardill, Irene (2002). *Gender, Migration and the Dual Career Household*. London and New York: Routledge.

Hardill, Irene & MacDonald, Sue (1998). Choosing to relocate: An examination of the impact of expatriate work on dual-career households. *Women's Studies International Forum*, 21: 21–29. http://dx.doi.org/10.1016/S0277-5395(97)00085-X

Harker, Christopher & Martin, Lauren (2012). Guest editorial. Familial relations: Spaces, subjects, and politics. *Environment and Planning A*, 44: 768–775. DOI: 10.1068/a4513

Harvey, William (2008). The social networks of British and Indian expatriate scientists in Boston. *Geoforum*, 39: 1756–1765. http://dx.doi.org/10.1016/j.geoforum.2008.06.006

Harrison, Edelweiss & Michailova, Snejina (2012). Working in the Middle East: Western female expatriates' experiences in the United Arab Emirates. *International Journal of Human Resource Management*, 23(4): 625–644. DOI: 10.1080/09585192. 2011.610970

Hart, John Fraser (2001). No dead rabbits. *The Geographical Review*, 91: 322–327. DOI: 10.1111/j.1931-0846.2001.tb00486.x

Heard-Bey, Frauke (2001). The tribal society of the UAE and its traditional economy. In Ibrahim Al-Abed & Peter Hellyer (eds), *The United Arab Emirates: A new perspective*. London: Trident Press, pp. 98–116.

Heard-Bey, Frauke (2004). *From Trucial States to United Arab Emirates: A society in Transition*. Dubai: Motivate Publishing.

Heath, Sue (2004). Shared households, quasi-communes and neo-tribes. *Current Sociology*, 52(2): 161–179. DOI: 10.1177/0011392104041799

Ho, Elaine (2006). Negotiating belonging and perceptions of citizenship in a transnational world: Singapore, a cosmopolis? *Social and Cultural Geography*, 7, 385–401. http://dx.doi.org/10.1080/14649360600715086

Hobbs, Dick & May, Tim (eds) (1993). *Interpreting the Field*. Oxford: Clarenden Press.

Hochschild, Arlie (1996). The emotional geography of work and family life. In Lydia Morris & E. Stina Lyon (eds), *Gender Relations in Public and Private*. Basingstoke: MacMillan, pp. 13–32.

Hochschild, Arlie (2003). *The Managed Heart: The Commercialization of Human Feeling*. London: University of California Press.

Hopkins, Peter & Noble, Greg (2009). Masculinities in place: Situated identities, relations and intersectionality. *Social and Cultural Geography*, 10: 811–819. http://dx.doi.org/10.1080/14649360903305817

Hoskins, Janet (1998). *Biographical Objects: How Things Tell the Stories of People's Lives*. London: Routledge.

Huang, Shirlena & Yeoh, Brenda (2008). Heterosexualities and the global(ising) city in Asia. *Asian Studies Review*, 32, 1–6. http://dx.doi.org/10.1080/10357820701871187

Hubbard, Phil (2002). Sexing the self: Geographies of engagement and encounter. *Social and Cultural Geography*, 3: 365–381. http://dx.doi.org/10.1080/1464936021000032478

Hubbard, Phil (2008). Here, there, everywhere: The ubiquitous geographies of heteronormativity. *Geography Compass*, 2, 1–19. DOI: 10.1111/j.1749-8198.2008.00096.x

Human Rights Watch (2006). Building towers, cheating workers: Exploitation of migrant construction workers in the United Arab Emirates. *Human Rights Watch*, 18(8). http://www.hrw.org/en/reports/2006/11/11/building-towers-cheating-workers

Hyam, Ronald (1990). *Empire and Sexuality: The British Experience*. Manchester: Manchester University Press.

Jackson, Peter, Crang, Philip & Dwyer, Claire (eds) (2004). *Transnational Spaces*. London: Routledge.

Jackson, Stevi (1999). *Heterosexuality in Question*. London: Sage.

Jamieson, Lynn (1998). *Intimacy: Personal Relationships in Modern Societies*. Cambridge: Polity Press.

Jayne, Mark, Holloway, Sarah & Valentine, Gill (2006). Drunk and disorderly: Alcohol, urban life and public space. *Progress in Human Geography*, 30(4): 451–468. https://doi.org/10.1191/0309132506ph618oa

Johnson, Mark (2015). Gendering pastoral power: Masculinity, affective labour and competitive bonds of solidarity among Filipino migrant men in Saudi Arabia. *Gender, Place and Culture.* DOI: 10.1080/0966369X.2015.1090411.

Johnston, Lynda & Longhurst, Robyn (2010). *Space, Place and Sex: Geographies of Sexualities.* Plymouth: Rowman and Littlefield Publishers.

Junemo, Mattias (2004). 'Let's build a palm island!': Playfulness in complex times. In Mimi Sheller & John Urry (eds), *Tourism Mobilities: Places to Play, Places in Play.* London: Routledge, pp. 181–190.

Kamrava, Mehran & Babar, Zahra (eds) (2012). *Migrant Labor in the Persian Gulf.* Center for International and Regional Studies, Georgetown University Qatar, and Hurst and Company London.

Kanna, Ahmed (2010). Flexible citizenship in Dubai: Neoliberal subjectivity in the emerging 'city-corporation'. *Cultural Anthropology,* 25: 100–129. DOI: 10.1111/j.1548-1360.2009.01053.x

Kanna, Ahmed (ed.) (2011). *Dubai: The City as Corporation.* Minneapolis: University of Minnesota Press.

Kanna, Ahmed (2014). A group of like-minded lads in heaven: Everydayness and the production of Dubai space. *Journal of Urban Affairs,* 36(S2): 605–620.

Kapiszewski, Andrzej (2001). *Nationals and Expatriates: Population and Labour Dilemmas of the Gulf Council States.* Reading: Garnet Publishing Ltd.

Kapiszewski, Andrzej (2003). The changing status of Arab migrant workers in the GCC. *Journal of Social Affairs,* 20(78): 33–60.

Kapiszewski, Andrzej (2006). Arab versus Asian migrant workers in the GCC countries. Retrieved from: http://www.un.org/esa/population/meetings/EGM_Ittmig_Arab/P02_Kapiszewski.pdf

Kaplan, Caren (1996). *Questions of Travel: Postmodern Discourses of Displacement.* London: Duke University Press.

Katz, Cindi (1994). Playing the field: Questions of fieldwork in geography. *Professional Geographer,* 46(1): 67–72. DOI: 10.1111/j.0033-0124.1994.00067.x

Kendall, David (2012). Always let the road decide: South Asian labourers along the highways of Dubai, UAE: A photographic essay. *South Asian Diaspora,* 4(1): 45–55. http://dx.doi.org/10.1080/19438192.2012.634561

Kenworthy Teather, Elizabeth (ed.) (1999). *Embodied Geographies: Spaces, Bodies and Rites of Passage.* London: Routledge.

Kipnis, Laura (2003). *Against Love: A Polemic.* New York: Vintage Books.

King, Russell (2002). Towards a new map of European migration. *International Journal of Population Geography,* 8(2), 889–106. DOI: 10.1002/ijpg.246

King, Russell, Warnes, Tony & Williams, Allan (2000). *Sunset Lives: British Retirement Migration to the Mediterranean.* Oxford: Berg Publishers.

Knight, Sara (2003). Disappearing boyfriends. *7Days,* Friday 15 August.

Knopp, Larry (2004). Ontologies of place, placelessness, and movement: Queer quests for identity and their impacts on contemporary geographic thought. *Gender, Place and Culture,* 11: 121–134. http://dx.doi.org/10.1080/0966369042000188585

Knowles, Caroline (2005). Making whiteness: British lifestyle migrants in Hong Kong. In Caroline Knowles & Claire Alexander (eds), *Making Race Matter: Bodies, Space and Identity.* Basingstoke: Palgrave Macmillan.

Knowles, Caroline & Harper, Douglas (2009). *Hong Kong: Migrant Lives, Landscapes and Journeys.* London: University of Chicago Press.

Kofman, Eleonore & Raghuram, Parvati (2005). Gender and skilled migrants: Into and beyond the work place. *Geoforum*, 36: 149–154. http://dx.doi.org/10.1016/j.geoforum.2004.06.001

Krane, Jim (2009). *Dubai: The Story of the World's Fastest City*. Atlantic Books, London.

Lamb, Sarah (2002). Intimacy in a transnational era: The remaking of aging among Indian Americans. *Diaspora*, 11(3): 299–330. DOI: 10.1353/dsp.2011.0003

Laurie, Nina, Dwyer, Claire, Holloway, Sara & Smith, Fiona (1999). *Geographies of New Femininities*. Harlow: Longman Pearson.

Leggett, William (2013). *The Flexible Imagination: At Work in the Transnational Corporate Offices of Jakarta, Indonesia*. Plymouth: Lexington Books.

Leonard, Pauline (2008). Migrating identities: Gender, whiteness and Britishness in post-colonial Hong Kong. *Gender, Place and Culture*, 15(1): 45–60. DOI: 10.1080/09663690701817519

Leonard, Pauline (2010). *Expatriate Identities in Postcolonial Organizations: Working Whiteness*. Hampshire: Ashgate.

Lester, Alan (2012). Foreword. In Anne-Meike Fechter & Katie Walsh (eds), *The New Expatriates: Postcolonial Approaches to Mobile Professionals*. Oxon: Routledge, pp. 1–8.

Levin, Irene (2004). Living apart together: A new family form. *Current Sociology*, 52(2): 223–240. http://dx.doi.org/10.1177/0011392104041809

Licoppe, Christian (2004). 'Connected' presence: The emergence of a new repertoire for managing social relationships in a changing communication technoscape. *Environment and Planning D: Society and Space*, 22(1): 135–156.

Lloyd, Jenny (2017). 'You're not big, you're just in Asia': Expatriate embodiment and emotional experiences of size in Singapore. *Social & Cultural Geography*, DOI: 10.1080/14649365.2017.1384047

Longhurst, Robyn (2001). *Bodies: Exploring Fluid Boundaries*. London: Routledge.

Luibhéid, Eithne (2005). Introduction: Queering migration and citizenship. In Eithne Luibhéid & Lionel Cantú Jr. (eds), *Queer Migrations: Sexuality, U.S. Citizenship, and Border Crossings*. London & Minneapolis: University of Minnesota Press, pp. ix–xlvi.

Luibhéid, Eithne & Cantú, Lionel Jr. (2005). *Queer Migrations: Sexuality, U.S. Citizenship, and Border Crossings*. London & Minneapolis: University of Minnesota Press.

Maclaren, Kym (2014). Intimacy and embodiment: An introduction. *Emotion, Space and Society*, 13: 55–64. http://dx.doi.org/10.1016/j.emospa.2014.09.002

Mahdavi, Pardis (2011). *Gridlock: Labor, Migration, and Human Trafficking in Dubai*. California: Stanford University Press.

Mahdavi, Pardis (2016). *Crossing the Gulf: Love and Family in Migrant Lives*. California: Stanford University Press.

Mai, Nicola & King, Russell (2009). Love, sexuality and migration: Mapping the issue(s). *Mobilities*, 4(3): 295–307.

Mains, Susan, Gilmartin, Mary, Cullen, Declan, Mohammad, Robina, Tolia-Kelly, Divya P., Raghuram, Parvati & Winders, Jamie (2013). Postcolonial migrations. *Social & Cultural Geography*, 14(2): 131–144. DOI: 10.1080/14649365.2012.753468.

Malam, Linda (2008). Bodies, beaches and bars: Negotiating heterosexual masculinity in southern Thailand's tourism industry. *Gender, Place and Culture* 15(6): 581–594. http://dx.doi.org/10.1080/09663690802518461.

Malecki, Edward J. & Ewers, Michael C. (2007). Labour migration to world cities: With a research agenda for the Arab Gulf. *Progress in Human Geography*, 31: 467–484. DOI: 10.1177/0309132507079501

Manalansan, Martin (2006). Queer intersections: Sexuality and gender in migration studies. *International Migration Review*, 40: 224–249. DOI: 10.1111/j.1747-7379.2006.00009.x

Mason, Jennifer (2004). Managing kinship over long distances: The significance of 'the visit'. *Social Policy and Society*, 3: 421–429. DOI: 10.1017/S1474746404002052

Massey, Doreen (1994). *Space, Place and Gender*. Minneapolis: University of Minnesota Press.

McKay, Deidre (2006). Translocal circulation: Place and subjectivity in an extended Filipino community. *The Asia Pacific Journal of Anthropology*, 7(3) 265–278, DOI: 10.1080/14442210600979357

McKay, Deidre (2007). 'Sending dollars shows feeling': emotions and economies in Filipino migration. *Mobilities*, 2(2): 175–194. DOI: 10.1080/17450100701381532

Mee, Kathleen & Wright, Sarah (2009). Guest editorial: Geographies of belonging. *Environment and Planning A*, 41: 772–779. DOI: 10.1068/a41364

Meier, Lars (ed.) (2015). *Migrant professionals in the city. Local encounters, identities and inequalities*. New York/London: Routledge.

Mohammad, Robina & Sidaway, James (2012). Spectacular urbanization amidst variegated geographies of globalization: Learning from Abu Dhabi's trajectory through the lives of South Asian men. *International Journal of Urban and Regional Research*, 36(3): 606–627. DOI: 10.1111/j.1468-2427.2011.01099.x

Mohammad, Robina & Sidaway, James (2016). Shards and stages: Migrant lives, power and space viewed from Doha, Qatar. *Annals of the Association of American Geographers*, 106(6): 1397–1417. 10.1080/24694452.2016.1209402

Morgan, David (1996). *Family Connections: An Introduction to Family Studies*. Cambridge: Polity Press.

Morgan, David (2013). *Rethinking Family Practices*. Basingstoke: Palgrave MacMillan.

Morley, David G. (ed.) (2000). *Home Territories: Media, Mobility and Identity*. London: Routledge.

Morrison, Carey-Ann (2012). Heterosexuality and home: Intimacies of space and spaces of touch. *Emotion, Space and Society*, 5: 10–18. http://dx.doi.org/10.1016/j.emospa.2010.09.001

Morrison, Carey-Ann (2013). Homemaking in New Zealand: Thinking through the mutually constitutive relationship between domestic material objects, heterosexuality and home. *Gender, Place and Culture* 20(4):413–431. DOI:10.1080/0966369X.2012.694358

Morrison, Carey-Ann, Johnston, Lynda & Longhurst, Robyn (2013). Critical geographies of love as spatial, relational and political. *Progress in Human Geography*, 37(4): 505–521. http://dx.doi.org/10.1177/0309132512462513

Moss, Pamela & Donovan, Courtney (eds) (2017). *Writing Intimacy into Feminist Geography*. Abingdon and New York: Routledge.

Nast, Heidi (1994). Women in the field: Critical feminist methodologies and theoretical perspectives. *The Professional Geographer*, 46: 54–66. DOI: 10.1111/j.0033-0124.1994.00054.x

Nielson, Niels (1999). Knowledge by doing: Home and identity in bodily perspective, in David Crouch (ed.), *Leisure/Tourism Geographies: Practices and Geographical Knowledge*. London: Routledge, pp. 277–289.

Oliver, Caroline (2008). *Retirement Migration: Paradoxes of Ageing*. New York/Abingdon: Routledge.

Onley, James (2005). Britain's informal empire in the Gulf, 1820–1971. *Journal of Social Affairs*, 22: 29–45.

O'Reilly, Karen (2000). *The British on the Costa Del Sol*. London: Routledge

Osella, Filippo & Osella, Caroline (2000). Migration, money and masculinity in migration to the Gulf. *Journal of the Royal Anthropological Institute*, 6: 117–133. DOI: 10.1111/1467-9655.t01-1-00007

Osella, Filippo & Osella, Caroline (2007). 'I am Gulf': The production of cosmopolitanism in Kozhikode, Kerala, India. In Edward Simpson & Kai Kresse (eds), *Struggling with History: Islam and Cosmopolitanism in the Western Indian Ocean*. London: C. Hurst & Co. Publishers Ltd, pp. 323–355.

Oswin, Natalie (2008). Critical geographies and the uses of sexuality: Deconstructing queer space. *Progress in Human Geography*, 32:89–103. DOI: 10.1177/0309132507085213.

Oswin, Natalie. (2010a). Sexual tensions in modernizing Singapore: The postcolonial and the intimate. *Environment and Planning D: Society and Space*, 28: 128–141. DOI: 10.1068/d15308.

Oswin, Natalie (2010b). The modern model family at home in Singapore: A queer geography. *Transactions of the Institute of British Geographers*, 35: 256–268. DOI: 10.1111/j.1475-5661.2009.00379.x

Oswin, Natalie & Olund, Eric (2010). Governing intimacy. *Theme issue of Environment and Planning D: Society and Space* 28(1): 60–67. DOI: 10.1068/d2801ed

Padilla, Mark, Hirsch, Jennifer, Munoz-Laboy, Miguel, Sember, Robert & Parker, Richard (eds) (2007). *Love and Globalization: Transformations of Intimacy in the Contemporary World*. Nashville: Venderbilt University Press.

Pahl, Ray (2000). *On Friendship*. London: Polity Press.

Pahl, Ray (2002). Towards a more significant sociology of friendship. *European Journal of Sociology*, 43: 410–423. http://dx.doi.org/10.1017/S0003975602001169

Pahl, Ray & Spencer, Liz (2004). Personal communities: Not simply families of 'fate' or 'choice'. *Current Sociology*, 52(2): 199–221.

Pain, Rachel (2009). Globalized fear? Towards an emotional geopolitics. *Progress in Human Geography*, 33(4): 466–486. http://dx.doi.org/10.1177/0309132508104994

Pain, Rachel & Staeheli, Lynn (2014). Introduction: Intimacy-geopolitics and violence. *Area*, 46(4): 344–360. DOI: 10.1111/area.12138

Parreñas, Rhacel (2001). Mothering from a distance: Emotions, gender and intergenerational relations in Filipino transnational families. *Feminist Studies*, 27(2): 361–390.

Parreñas, Rhacel (2005). *Children of Global Migration: Transnational Families and Gendered Woes*. Stanford: Stanford University Press.

Pearson, Geoffrey (1993). Foreword. Talking a good fight: Authenticity and distance in the ethnographer's craft. In Dick Hobbs & Tim May (eds), *Interpreting the Field*. Oxford: Clarendon Press, pp. vii–xx.

Pile, Steve (2010). Emotions and affect in recent human geography. *Transactions of the Institute of British Geographers*, 35(1): 5–20. http://dx.doi.org/10.1111/j.1475-5661.2009.00368.x

Piper, Nicola & Roces, Mina (eds) (2003). *Wife or Worker? Asian Women and Migration*. Oxford: Rowman and Littlefield Publishers, Inc.

Povinelli, Elizabeth (2006). *The Empire of Love: Toward a Theory of Intimacy, Genealogy, and Carnality*. Durham, NC: Duke University Press.

Pratt, Geraldine (2012). *Families Apart: Migrant Mothers and the Conflicts of Labor and Love*. Minneapolis: University of Minnesota Press.

Pratt, Geraldine & Rosner, Victoria (eds) (2012). *The Global and the Intimate: Feminism in our Time*. New York: Columbia University Press.

Price, Marie & Benton-Short, Lisa (2007). Immigrants and world cities: From the hyper-diverse to the bypassed. *Geojournal*, 68: 103–117. DOI: 10.1007/s10708-007-9076-x

Puar, Jasbir K. (2006): Mapping US homonormativities. *Gender, Place and Culture*, 13(1): 67–88. http://dx.doi.org/10.1080/09663690500531014

Rawlins, William (1992). Friendship Matters: Communication, Dialectics, and the Life Course. Hawthorne, NY: Aldine.

Redman, Peter (2002). Love is in the air: Romance and the everyday. In Tony Bennett & Diane Watson (eds), *Understanding Everyday Life*. Oxford: Blackwell, pp. 51–90.

Reynolds, Jill (2008). *The Single Woman*. London: Routledge.

Riemsdijk, Micheline van (2014). International migration and local emplacement: Everyday place-making practices of skilled migrants in Oslo, Norway. *Environment and Planning A*, 46(4): 963–979. http://dx.doi.org/10.1068/a46234

Robinson, Victoria, Hockey, Jenny & Meah, Angela (2004). What I used to do ... on my mother's settee': Spatial and emotional aspects of heterosexuality in England. *Gender, Place and Culture*, 11(3): 417–435. DOI: 10.1080/0966369042000258712

Rose, Gillian (1995). Geography and gender, cartographies and corporealities. *Progress in Human Geography*, 19: 544–48. http://dx.doi.org/10.1177/030913259501900407

Rose, Gillian (2003). Family photographs and domestic spacings: A case study. *Transactions of the Institute of British Geographers*, 28: 5–18. DOI: 10.1111/1475-5661.00074

Rose, Gillian (2004). Situating knowledges: Positionality, reflexivities and other tactics. In Nigel Thrift & Sarah Whatmore (eds), *Cultural Geography: Critical Concepts in the Social Sciences*. London: Routledge, pp. 244–262.

Roseneil, Sasha & Budgeon, Shelley (2004). Cultures of intimacy and care beyond 'the family': Personal life and social change in the early 21st century. *Current Sociology*, 52: 135–159. http://dx.doi.org/10.1177/0011392104041798

Roy, Ananya (2009). The 21st-century metropolis: New geographies of theory. *Regional Studies*, 43(6): 819–830. DOI: 10.1080/00343400701809665

Ryan, Louise & Mulholland, Jon (2014). Trading places: French highly skilled migrants negotiating mobility and emplacement in London. *Journal of Ethnic and Migration Studies*, 40: 584–600. http://dx.doi.org/10.1080/1369183X.2013.787514

Sabban, Rima (2004). Women migrant domestic workers in the United Arab Emirates. In Simel Esim & Monica Smith (eds), *Gender and Migration in Arab States: The Case of Domestic Workers*. Beirut: International Labour Organization, pp. 86–108.

Said, Edward (2001). *Orientalism: Western Conceptions of the Orient*. London: Penguin.

Said, Edward (2003). *Orientalism*. London: Penguin.

Sanchez Taylor, Jacqueline (2000). Tourism and 'embodied' commodities: Sex tourism in the Caribbean. In Stephen Clift & Simon Carter (eds), *Tourism and Sex: Culture, Commerce and Coercion*. London: Pinter, pp. 41–53.

Schlote, Christiane (2014). Writing Dubai: Indian labour migrants and taxi topographies. *South Asian Diaspora*, 6(1): 33–46. http://dx.doi.org/10.1080/19438192.2013.828500

Scott, Sam (2006). The social morphology of skilled migration: The case of the British middle class in Paris. *Journal of Ethnic and Migration Studies*, 32(7): 1105–1129. http://dx.doi.org/10.1080/13691830600821802

Scott, Sam (2007). The community morphology of skilled migration: The changing role of voluntary and community organisations (VCOs) in the grounding of British identities in Paris (France). *Geoforum*, 38: 655–676. http://dx.doi.org/10.1016/j.geoforum.2006.11.015

Seabrook, Jeremy (1996). *Travels in the Skin Trade: Tourism and the Sex Industry*. London: Pluto.

Sedgwick, Eve (1999). *A Dialogue on Love*. Boston: Beacon Press.

Shah, Nasra (2008). Recent labor immigration policies in the oil-rich Gulf: How effective are they likely to be? *ILO Asian Regional Programme on Governance of Labour Migration Working Paper No. 3*. http://digitalcommons.ilr.cornell.edu/intl/52

Sheller, Mimi & Urry, John (eds) (2008). *Tourism Mobilities: Places to Play, Places in Play*. London: Routledge.

Shen, Hsui-hua (2008). The Purchase of transnational intimacy: Women's bodies, transnational masculine privileges in Chinese economic zones. *Asian Studies Review*, 32: 57–75. http://dx.doi.org/10.1080/10357820701870759

Silverstone, Roger & Hirsch, Eric (1992). *Consuming Technologies: Media and Information in Domestic Spaces*. London and New York: Routledge.

Silvey, Rachel (2003). Gender and mobility: Critical ethnographies of migration in Indonesia. In Alison Blunt, Pyrs Gruffudd, Jon May, Miles Ogborn & David Pinder (eds), *Cultural Geography in Practice*. London: Arnold, pp. 91–105.

Silvey, Rachel (2005). Borders, embodiment and mobility: Feminist migration studies in geography. In Lise Nelson & Joni Seager (eds), *A Companion to Feminist Geography*. Malden MA: Wiley InterScience, Blackwell, pp. 138–149.

Silvey, Rachel (2007). Mobilizing piety: Gendered morality and Indonesian–Saudi transnational migration. *Mobilities*, 2(2): 219–229. DOI: 10.1080/17450100701381565

Silvey, Rachel & Lawson, Victoria (1999). Placing the migrant. *Annals of the Association of American Geographers*, 89: 121–132. DOI: 10.1111/0004-5608.00134

Sklair, Leslie (2000). *The Transnational Capitalist Class*. Oxford: Wiley Blackwell.

Skrbiš, Zlatko (2008). Transnational families: Theorising migration, emotions and belonging. *Journal of Intercultural Studies*, 29(3): 231–246. DOI: 10.1080/07256860802169188

Smart, Carol (2007). *Personal Life: New Directions in Sociological Thinking*. Cambridge: Polity Press.

Smart, Carol & Neale, Bren (1999). *Family Fragments?* Cambridge: Polity Press.

Smith, Benjamin (2010). Scared by, of, in, and for Dubai. *Social and Cultural Geography*, 11(3): 263–282. DOI: 10.1080/14649361003637182

Smith, Michael Peter (2000). *Transnational Urbanism: Locating Globaliztion*. Oxford: Blackwell.

Smith, Michael Peter (2005). Transnational urbanism revisited. *Journal of Ethnic and Migration Studies*, 31: 235–244. DOI: 10.1080/1369183042000339909

Smith, Sara (2016). Intimacy and angst in the field. *Gender, Place and Culture*, 23(1): 134–146. DOI: 10.1080/0966369.2014.958067

Spencer, Lisa & Pahl, Ray (eds) (2006). *Rethinking Friendship: Hidden Solidarities Today*. Princeton, NJ: Princeton University Press.

Sriskandarajah, Danny & Drew, Catherine (2006). *Brits Abroad: Mapping the Scale and Nature of British Emigration*. London: Institute for Public Policy Research.

Stanley, Liz & Wise, Sue (1993). *Breaking Out Again: Feminist Ontology and Epistemology*. London: Routledge.

Stephenson, Marcus (2013). Tourism, development and 'destination Dubai': Cultural dilemmas and future challenges. *Current Issues in Tourism*, 17(8): 723–738.

Svašek, Maruska (2010). On the move: Emotions and human mobility. *Journal of Ethnic and Migration Studies*, 36(6): 865–880. http://dx.doi.org/10.1080/13691831003643322

Thien, Deborah (2005). Intimate distances: Considering questions of 'us'. In Joyce Davidson, Liz Bondi & Mick Smith (eds), *Emotional Geographies*. Aldershot: Ashgate, pp. 191–204.

Thomas, Mandy (1999). *Dreams in the Shadows: Vietnamese-Australian Lives in Transition*. Australia: Allen and Unwin.

Thrift, Nigel (2005). But malice aforethought: Cities and the natural history of hatred. *Transactions of the Institute of British Geographers*, NS30: 133–150. DOI: 10.1111/j.1475-5661.2005.00157.x

Till, Karen (2001). Returning home and to the field. *The Geographical Review*, 91: 46–56. DOI: 10.1111/j.1931-0846.2001.tb00457.x

Urry, John (2004). Connections. *Environment and Planning D: Society and Space*, 22: 27–37. DOI: 10.1068/d322t

Valenta, Marko & Jakobsen, Jo (2016). Moving to the Gulf: An empirical analysis of the patterns and drivers of migration to the GCC countries, 1960–2013. *Labor History*, 57(5): 627–648. http://dx.doi.org/10.1080/0023656X.2016.1239885

Valentine, Gill (1993). Desperately seeking Susan: A geography of lesbian friendships. *Area*, 25(2): 109–116. http://www.jstor.org/stable/20003237

Valentine, Gill (1999). Doing household research: Interviewing couples together and apart. *Area*, 31: 67–74. DOI: 10.1111/j.1475-4762.1999.tb00172.x

Valentine, Gill (2001). *Social Geographies: Space and Society*. London: Prentice Hall.

Valentine, Gill (2008). The ties that bind: Towards geographies of intimacy. *Geography Compass*, 2(6): 2097–2110. DOI: 10.1111/j.1749-8198.2008.00158.x

Valentine, Gill & Hughes, Kathryn (2012). Shared space, distant lives? Understanding family and intimacy at home through the lens of internet gambling. *Transactions of the Institute of British Geographers*, 37: 242–255. DOI: 10.1111/j.1475-5661.2011.00469.x

Van Hoven, Bettina & Hörschelmann, Kathryn (eds) (2005). *Spaces of Masculinities*. London: Routledge.

Vertovec, Steven (2001). Transnationalism and identity. *Journal of Ethnic and Migration Studies*, 27(4): 573–582. http://dx.doi.org/10.1080/13691830120090386

Vora, Neha (2008). Producing diasporas and globalization: Indian middle-class migrants in Dubai. *Anthropological Quarterly*, 81(2): 377–406. DOI: 10.1353/anq.0.0010

Vora, Neha (2013). *Impossible Citizens: Dubai's Indian Diaspora*. Durham and London: Duke University Press.

Waitt, Gordon & Gorman-Murray, Andrew (2011). 'It's about time you came out': Sexualities, mobility and home. *Antipode*, 43: 1380–1403. DOI: 10.1111/j.1467-8330.2011.00876.x

Walsh, Katie (2006a). 'Dad says I'm tied to a shooting star!' Grounding (research on) British expatriate belonging. *Area*, 38: 268–278. DOI: 10.1111/j.1475-4762.2006.00687.x

Walsh, Katie (2006b). British expatriate belongings: Mobile homes and transnational homing. *Home Cultures*, 3: 123–144.

Walsh, Katie (2007). 'It got very debauched, very Dubai!' Performances of heterosexuality amongst single British expatriates. *Social and Cultural Geography*, 8(4): 507–535.

Walsh, Katie (2008). Travelling together? Work, intimacy and home amongst British expatriate couples in Dubai. In Anne Coles & Anne-Meike Fechter (eds), *Beyond the Incorporated Wife: Gender Relations among Mobile Professionals*. London: Routledge, pp. 63–84.

Walsh, Katie (2009). Geographies of the heart in transnational spaces: Love and the intimate lives of British migrants in Dubai. *Mobilities* (Special Issue), 4: 427–445. http://dx.doi.org/10.1080/17450100903195656

Walsh, Katie (2010). Negotiating migrant status in the emerging global city: Britons in Dubai. *Encounters*, 2: 235–255. ISSN 2075-048X

Walsh, Katie (2011). Migrant masculinities and domestic space: British home-making practices in Dubai. *Transactions of the Institute of British Geographers*, 36 (4): 516–529. DOI: 10.1111/j.1475-5661.2011.00442.x

Walsh, Katie (2012). Emotion and migration: British transnationals in Dubai. *Environment and Planning D: Society and Space*, 30(1): 43–59. DOI: http://dx.doi.org/10.1068/d12409

Walsh, Katie (2014). Placing transnational migrants through comparative research: British migrant belonging in five GCC cities. *Population, Space and Place*, 20 (1): 1–17. ISSN 1544-8444

Walsh, Katie, Shen, Hsiu-hua & Willis, Katie (2008). Introduction: Heterosexuality and migration in Asia. *Gender, Place and Culture* (Special Issue), 15: 575–579. http://dx.doi.org/10.1080/09663690802518438

Weston, Kath (1991). *Families We Choose: Lesbians, Gays and Kinship*. New York and London: Columbia University Press.

Whitehead, Stephen (2002). *Men and Masculinities*. Cambridge: Polity Press.

Wilkinson, Eleanor (2014). Single people's geographies of home: Intimacy and friendship beyond 'the family'. *Environment and Planning A*, 46(10): 2452–2468. DOI: 10.1068/a130069p

Williams, Jeremy (1998). *Don't They Know it's Friday? Cross-cultural Considerations for Business and Life in the Gulf*. UAE: Motivate.

Willis, Katie & Yeoh, Brenda (2002). Gendering transnational communities: A comparison of Singapore and British migrants in China. *Geoforum*, 33: 553–565. http://dx.doi.org/10.1016/S0016-7185(02)00039-8

Willis, Katie & Yeoh, Brenda (2003). Gender, marriage and skilled migration: The case of Singaporeans in China. In Nicola Piper & Mina Roces (eds), *Wife or Worker? Asian Women and Migration*. Oxford: Rowman and Littlefield Publishers, Inc., pp. 101–120.

Willis, Katie & Yeoh, Brenda (2007). 'Coming to China changed my life': Gender roles and relations among single British migrants. In Anne Coles & Anne-Meike Fechter (eds), *Gender and Family among Transnational Professionals*. London: Routledge, pp. 211–232.

Wilson, Ara (2012). Intimacy: A useful category of transnational analysis. In Geraldine Pratt & Victoria Rosner (eds), *The Global and the Intimate: Feminism in Our Time*. New York: Columbia University Press, pp. 31–56.

Yeoh, Brenda (1999). Global/globalizing cities. *Progress in Human Geography*, 23(4): 607–616. http://dx.doi.org/10.1191/030913299674647857

Yeoh, Brenda (2001). Postcolonial cities. *Progress in Human Geography*, 25(3): 456–468. http://dx.doi.org/10.1191/030913201680191781

Yeoh, Brenda & Huang, Shirlena (2010). Sexualized politics of proximities among female transnational migrants in Singapore. *Population, Space and Place*, 16: 37–49. DOI: 10.1002/psp.579

Yeoh, Brenda & Willis, Katie (2005). Singaporean and British transmigrants in China and the cultural politics of 'contact zones'. *Journal of Ethnic and Migration Studies*, 31(2): 269–285. http://dx.doi.org/10.1080/1369183042000339927

Yeoh, Brenda, Willis, Katie & S. M. Abdul Khader Fakhri (2003). Introduction: Transnationalism and its edges. *Ethnic and Racial Studies*, 26: 207–217. DOI: 10.1080/0141987032000054394

Index

Transnational Geographies of the Heart: Intimate Subjectivities in a Globalising City, First Edition. Katie Walsh.
© 2018 John Wiley & Sons Ltd. Published 2018 by John Wiley & Sons Ltd.

heteronormativity (*cont'd*)
legal, 108–110, 125, 147
and sexuality, 38, 107–108
Hirsch, Eric, 103
Hochschild, Arlie, 27, 134
home, 126
attachment, 113–114, 139–140,
141, 142
concept, 28, 30, 37, 103, 136, 138–139
critical geographies, 38–39, 126
domestic artefacts, 136–137
and family, 126, 130, 136–138,
152–153
ideology of, 126
see also homemaking; shared households
homemaking, 20, 64, 132, 133, 136–138
feminised *see* domesticity
Hong Kong, 6, 12, 85, 90–91
Huang, Shirlena, 110
Hubbard, Phil, 31
human rights
and low-income migrants, 7, 54
reproductive, 4

Indian migrants, 53, 54–56, 84
as quasi-citizens, 72
individualisation theories, 33–35
Indonesia, 12, 96
intimacies, personal, 3–4, 13–18, 146
and the body, 39–40, 145
caregiving, 27, 126
commodification of, 4
couple, 32, 33, 124, 141, 149: *see also*
couple ties
cross-cultural, 4, 66–67, 151
definitions, 17–18, 34, 41–42
families *see* families
friendship *see* friendship
geography of, 23–26, 30–31, 145
parent/child *see* parent–child
relationships
politicisation, 19
range, 4
romantic *see* romantic relationships
of the self, 145
sexual *see* sexual intimacies

sociology of, 33–37, 42–44
spatialisation, 15, 23, 37–38, 66, 143,
147–149
see also emotional labour; intimate
subjectivities; singleness
Institute of Public Policy Research,
5, 21n1
intimate subjectivities
expatriate, 102–125
terminology, 1, 4, 18, 37–44
Islam, 8, 11
and UAE citizenship, 51

Jackson, Stevi, 123
Jamieson, Lynn, 18, 19, 34, 40, 145
Johnston, Linda, 33
Jumeirah Beach Residence, Dubai,
150, 151

Kafala sponsorship system, 2, 7, 18, 38,
50–53, 56
and personal intimacies, 98–99
rationale, 111
regulatory framework, 11
residency applications, 74
Kamrava, Mehran, 50
Kanna, Ahmed, 10–11, 77
Kapiszewski, Andrzej, 53, 54, 56
Kendall, David, 82
King, Russell, 16, 22n5
kin support systems, 35, 101–102, 140,
142–143
and belonging, 25, 127
friendship substitutes, 101–102, 105
and technology, 144
Knowles, Caroline, 12
Kofman, Eleonore, 130
Krane, Jim, 8

labour migration, 4
agencies, 51
'bachelors', 51–52, 65, 115, 147
British *see* British migration
camps, 52, 55
domestic, 20: *see also* domestic
workers